NAVAL DOCKYARDS SOCIETY
Exploring the civil branches of navies and their

www.navaldockyards.org
Facebook: Naval-Dockyards-Society
Twitter @ navdocksoc
Supported by the National Maritime Museum, Greenwich

The Inaugural meeting of the Naval Dockyards Society was held at the National Maritime Museum, 14 September 1996 and the Constitutional meeting held at the National Maritime Museum on 1 March 1997.

Aims

1. Compile and distribute a list of members and their interests.
2. Publish two newsletters a year containing information, members' reports of activities, issues and book reviews.
3. Organise meetings and other relevant activities.
4. Increase public awareness of historic dockyards and related sites.
5. Increase access to historic dockyards and related sites.
6. Create links with related organisations in Britain and abroad.
7. Coordinate and promote new research.
8. Create a database of the present status of historic dockyards and related sites worldwide.
9. Offer assistance to those establishing dockyard heritage sites.
10. Encourage the storage and collection of oral history interviews related to dockyard history.
11. Coordinate the historical, architectural and technical expertise available within the society.
12. Compile a dockyards bibliography.

Navy Board Research Project

In 1999 the Society agreed to continue a project begun as a research undertaking by Richard Saville and Susan Lumas in the 1970s to create an index and database of class ADM 106 documents at the Public Record Office (now The National Archives). This collection of Navy Board in-letters 1658-1837 comprises miscellaneous correspondence from dockyard commissioners, officers and workers, naval captains and contractors from around the world. The project will foliate letters from ADM 106 and ADM B and ADM BP at the National Maritime Museum Greenwich, list and index their contents and input data to an on-line research resource. Researchers may access and interrogate the database electronically on The National Archives' online Catalogue.

Some 175,000 letters have been completed: 42% of the combined total of ADM 106 at Kew and of ADM B and ADM BP at Greenwich. Years available online are 1696, 1738-1741, 1744, 1745, 1750 and 1754-1770. Some of ADM B and ADM BP (listed as ADM 354) cover 1738-1739 and 1749-1758. The gap 1741-1744 represents damaged letters which are being conserved at Kew.

Fourteen volunteers work across the two sites. Among such routine matters as stores, transports, superannuation, surveys, embezzlement and wages, are unique items such as the ropemakers' 1675 mutiny, experiments and inventions, materials and medicines, Plymouth Yard expansion in 1761 and the Surveyor General of HM Woods' difficulties in America. This is an invaluable resource for international students of naval, dockyard, technological, social and administrative history.

The Society also supports Project Dockyard 2010: The History of Simon's Town Dockyard, 1900-2010

The South Africa Naval Heritage Trust and Simon's Town Historical Society are producing a history of Simon's Town Dockyard (East Yard) in 2010 to commemorate the centenary of commissioning the Selbourne Graving Dock. A Project Officer is collecting information from retired naval officers and 'Home Agreement' men. Contact Project Officer, Project Dockyard 2010, SA Naval Museum, Private Bag XI, Simon's Town, 7995, Republic of South Africa. kemndine1@telkomsa.net www.simonstown.com/stdc/civic/sths/dockyard2010.htm

Society Membership

Members receive a list of members' interests, two newsletters a year, with an annual conference presenting the latest research, published as *Transactions of the Naval Dockyards Society*, and the opportunity to participate in the research project, the Annual General Meeting, the Annual Conference and Tour. Currently we have around 200 members throughout the world: ex-dockyard personnel, family historians and academics.

Subscription rates from 1 January 2013

£25 *Individual Members*
£30 *Joint Members: two people living at one address*
£40 *Institutional Members*

Members are encouraged to pay by Standing Order. For overseas members (without a sterling bank account) to generate no exchange expenses or other deductions to the Society, the cheapest method is to pay by international money order. Our financial year begins on 1 January. Subscriptions may now be paid by PayPal. Look on our website.

For further details please contact the Hon. Secretary Dr Ann Coats.

2012-13 COMMITTEE

Hon. Chairman: Dr J D Davies
    'Borthwick', 18 Northill Road, Ickwell, Bedfordshire SG18 9ED

Hon. Secretary: Dr Ann Coats
    44 Lindley Avenue, Southsea PO4 9NU. Tel: 023 92863799. Email: ann.coats@port.ac.uk

Dockyards Editor (pro tem): Dr J D Davies

Hon. Treasurer: David Jenkins

Hon. Webmaster (pro tem): Dr J D Davies

Hon. Distribution Officer: David Baynes

Hon. Committee Members: David Baynes, Nicholas Blake, Dr Celia Clark, Steven Gray

Hon. Tours Organiser: David Baynes

Transactions Indexer: Andy Page

Ex Officio

Membership Secretary: Judith Webberley, Kimberley House, Old Mill Lane, Denmead, PO7 6JW. Tel: 023 9263 2645. Email: judith.nds@gmail.com

Navy Board Project Coordinator: Susan Lumas

Transactions Editor: Dr. Philip MacDougall

Auditor: Brian Vale

The Fourteenth Annual Conference, 2010
Theme: Pepys and Chips, Dockyards, Naval Administration and Warfare
in the Seventeenth Century

CONTENTS

	Page
Acknowledgements	5
Preface By RAY RILEY	7
English Naval Administration under Charles I – Top-down and Bottom-up – Tracing Continuities By ANN COATS	9
Parliament, Royal Dockyards and the London Maritime Community: the Aftermath of the 1648 Naval Revolt By RICHARD BLAKEMORE	31
Charles James and the Recreation of the Royal Navy 1660-1665 By HILARY TODD	45
The *London* of 1656: Her History and Armament By FRANK L FOX	57
The Women of Restoration Deptford By RICHARD ENDSOR	77
Intersections of Interest; A Prosopographical Analysis of Restoration Privateering Enterprise By RICHARD BRABANDER	87
Chatham to Erith via Dover. Charles II's Secret Foreign Policy and the Project for New Royal Dockyards, 1667-1672 By J D DAVIES	113

ACKNOWLEDGEMENTS

Ann Coats

This volume contains an exceptional range of remarkable images thanks to the efforts of our authors unearthing appropriate pictures, and to the generosity of the National Maritime Museum, the Trustees of the British Museum, the National Galleries of Scotland, Atlas Van Stolk, Rotterdam, The National Archives, Nico Brinck and Richard Endsor for supplying images free of copyright reproduction charges, and Bexley Archives and The National Archives of the UK for permission to reproduce documents.

Following our pattern of collaborating with other institutions to widen the NDS audience, in 2010 the Chair, Dr David Davies was pleased to welcome the Chair of the Samuel Pepys Club, Mrs Ann Sweeney, whom he had met in 2009 when the Club conferred upon him the fourth Samuel Pepys Award for *Pepys's Navy: Ships, Men and Warfare 1649-1689*.

Ann Sweeney reminded the conference in her welcome address that 2010 marked 350 years since Pepys had traveled to Breda with his patron Edward Mountagu to escort Charles II home to England. It was also the year he began his notable job as Clerk of the Acts. His famous *Diary* recorded his excitement at taking part in these unparalleled events and its minutiae make it a fascinating personal account.

Ann traced the development of the Samuel Pepys Club from the initial dinner at the Garrick Club on the bicentenary of Pepys's death on 26 May 1903, attended by Henry B Wheatley, editor of the most complete version of the *Diary of Samuel Pepys* (1893-1899) to date, Sir Frederick Bridge, Mr D'Arcy Power and Mr George Whale. They resolved that such a club would meet once a year to hear papers about Pepys and dine together. The first official Dinner was held on 1 December 1903 in Clothworkers' Hall. In 1913 the Club commemorated the 250th anniversary of the Royal Society, which Pepys joined in 1665 and was its President 1684-86, and will be celebrating the 350th anniversary in 2013. Dinners ceased for a while from 1914.

Regular activities include an annual memorial service near the date of his death in St Olave's, Hart Street, Pepys's burial place and the church of the Navy Board. Dinners or lunches are held at sites associated with Pepys: Magdalene College to see his books and presses, Windsor Castle to see Samuel Cooper's miniatures, including that of Elizabeth Pepys, Eton, Chatham, Hinchinbrooke and Brampton. Membership was set first at 50, then 70, his age at death. It is now limited to 140 UK and 14 overseas members, with institutions associated with Pepys, such as Trinity House, the Clothworkers' Company, Magdalene College and Charles Hoare Bank, as honorary members. It is an inclusive club with one purpose: 'to do honour to the memory of Samuel Pepys'.

Dr Davies thanked Ann Sweeney for her address and papers were then read. The Society is extremely grateful for the quality of their pieces, which were uniformly excellent. They also satisfied another objective of attracting a 'young guard' of new researchers, one of whom also demonstrated the appeal of the seventeenth century to the large American academic community. Every cloud has a silver lining. Richard Brabander could not return to the USA after the conference as planned, because the volcanic ash cloud prevented flights, so he took advantage of this enforced stay to carry out further PhD research at The National Archives. We are also indebted to Frank Fox for contributing his latest research to this volume. He did not attend the conference, but the subject is pertinent and its completion opportune.

A further aim of the NDS is to fulfill our remit to members who cannot attend. One member 'found the presentations excellent and stimulating. In fact I don't know where to jump first as far as reading goes and wonder if you would consider a mini-bibliography – two titles pertinent to their subject – while awaiting the published papers'. This proposal will be taken up following subsequent conferences, helping absent members engage with the themes.

Finally, in the Society's fifteenth year, this splendid volume of *Transactions* is testament to the skill, dedication and patience of editor Professor Ray Riley and Southsea printers Printcraft. This is the last volume to be edited by Professor Riley; the Society is greatly indebted to the time and effort he has contributed since 2004. The *Transactions* have been vital in raising the Society's profile internationally and this is clearly due to the quality displayed in these eight volumes. We are very grateful to him for carrying this rôle forward.

23 April 2012

PREFACE

Ray Riley

Since this volume of *Transactions* represents the fifteenth anniversary of conference papers, the committee considered it fitting that additional funds should be made available for colour photographs. Unfortunately the size of the issue, coupled with a very substantial rise in the cost of paper, have pushed the price even without colour to over £1000 more than the charge for volume 7. Regrettably therefore the inclusion of colour photographs cannot be justified.

Past issues of *Transactions* have tended to reflect particular aspects of the Society's interests: Gibraltar, Malta, Nelsonian Portsmouth, naval medicine, dockyard construction and wooden shipbuilding. The present volume has a much less specific theme, rather it focuses upon a period of time - the seventeenth century, resulting in a wide range of contributions. Perhaps not surprisingly, given the wealth of material he has bequeathed historians, Pepys surfaces in many of the papers.

Employing a top-down bottom-up framework, the underlying thrust of which endeavours to show that ultimate actions are a result of interpretation of decisions made by those at the topmost level of a hierarchy by those at a less exalted position, Ann Coats analyses the interaction between the Admiralty on the one hand and the decision-taking Navy Board on the other in the 1630s. Despite the changes which were effected, she argues that there were nevertheless continuities to be found – an issue easily overlooked. The role of the royal dockyards in the second civil war of 1648 is examined by Richard Blakemore; the contest for control of these facilities and the importance of the loyalty of dockyardmen, and indeed maritime communities, represented a key element in the struggle between royalists and parliamentarians. The plaudits for the recreation of the Navy in the 1660s are usually traced to Charles II, backed by Pepys, but Hilary Todd demonstrates the vital and very practical contribution made by James, Duke of York, brother of Charles II, as Lord High Admiral. Not a drinker, gambler or politician, his considerable administrative flair went unnoticed, Macaulay commenting that he would have made a respectable clerk in a dockyard.

From broad sweeps to the specific, Frank Fox investigates the circumstances surrounding the explosion in the *London* in 1665 in the Thames, killing almost all on board. Having done that, he endeavours to establish the nature of her armament and lists the fate of 41 of her guns. The article is, inter alia, a contribution to maritime archaeology. A second specific paper is that by Richard Endsor, who, in researching the Deptford ship the *Lenox*, noted several references to the feminine gender. He selects eight ladies who, in a variety of ways, had some impact on the yard. Almost inevitably, one had a relationship with Pepys. Richard Blakemore is concerned with privateering (not to be confused with piracy) which was a regulated business activity which not only benefited private investors, but also the state which was able to use the money generated to underpin the Army and the Navy. Pepys was involved, placing him in the possibly conflicting position of being both an investor and a public servant. A concluding piece by David Davies combines the general with the specific, examining the background to the secret Treaty of Dover, 1670. Among the least researched aspects of the Treaty was the proposal to acquire territory in the Netherlands, and to construct a new royal dockyard at Erith.

To conclude, Ann Coats has reminded me of the issue of dates relevant to this period. All dates are given in the Old Style, following the Julian Calendar, but the New Year is taken to begin on 1 January, with brackets modifying 1 January to 25 March, when the Julian New Year began. Where the original document acknowledged the New Year (for example 1665/6) this has been retained.

ENGLISH NAVAL ADMINISTRATION UNDER CHARLES I – TOP-DOWN AND BOTTOM-UP – TRACING CONTINUITIES*

Ann Coats

Abstract

Charles I's policies launched the Civil Wars in 1642, leading to his execution in 1649, a republic until 1660, when his son Charles II was restored, the succession of his second son James II in 1685 until his abdication in 1688, and the constitutional crowning of his grand-nephew William III and grand-daughter Mary II in 1689. This article gives an overview of top-down naval administration during the reigns of James I and Charles I, particularly abuses and reforms, and analyses bottom-up Navy Board correspondence to the Admiralty in the 1630s. The restoration of Charles II incorporated administrative continuities from both the earlier reigns and the interregnum.

Introduction

Naval administration under James I and Charles I set the context for the restoration naval administration. 1660s novice clerk of the acts Samuel Pepys sought validation from past practices through his position as record keeper and retriever. In 1662 and 1669 he made copies of the 1638 and 1659 *Discourses* of John Hollond and used them as a model: 'they hitting the very diseases of the Navy which we are troubled with nowadays', and 'making my people' read them. Hollond was navy surveyor 1649-52, during the interregnum, but he began as a clerk to assistant navy commissioner Captain Joshua Downing at Chatham in 1624. He progressed to navy paymaster 1635-*c*.1641 and navy commissioner *c*.1643-45.[1] Pepys started his career as clerk to George Downing, teller of the receipt at the Exchequer, *c*.1657, later Sir George Downing, 1st Baronet. Pepys distinguished disparagingly between 'what was before by proper distributions charged upon particular members', and during the interregnum 'committed to the management of the whole promiscuously'.[2] Hollond was not unique in publicising naval abuses, but his *Discourses* were promoted by restoration navy commissioner Admiral Sir William Penn and Pepys, who recognised continuing abuses and solutions which could be applied to the restoration administration.

Despite Hollond's condemnation of fraud, it was noted by Tanner that as navy paymaster he was accused in 1636 of paying seamen's tickets to absent parties (against the rules) for a 'gratification' and did not refute this. Another accusation was of charging commission of 2s in the pound before passing suppliers' payments for clothes from seamen's wages. Treasurer Sir William Russell defended him as acting on his instructions, but he was not free from some

Fig.1. Bust of Charles I, High Street, Old Portsmouth. Inscription reads: 'After his travels through all France into Spain and having paffed very many dangers both by sea and land he arrived here the 5th day of October, 1623'. Ann Coats, 2012.

---

*Paper presented at the fourteenth annual conference of the Naval Dockyards Society held at the National Maritime Museum, Greenwich, on 17 April 2010. Theme: Pepys and Chips, Dockyards, Naval Administration and Warfare in the Seventeenth Century.

of the abuses he criticised, as indeed neither was Pepys.[3]

When approaching the period 1600-40 it is impossible to forget that Charles I launched the Civil Wars in 1642 and was tried and executed by Parliament in 1649. Aylmer, aiming to pinpoint the development of the 'modern Civil Service' and providing invaluable detail, was himself informed by hindsight.[4] Charles's execution inaugurated a republic until 1660, when his son Charles II was restored, the succession of his second son James II in 1685 until his abdication in 1688, and the constitutional crowning of his grand-nephew William III and grand-daughter Mary II in 1689. To preserve the quality and loyalty of the dockyard workforce in 1660, Lord High Admiral James retained the increased salary rates of the Commonwealth, 'considering as well the greatnes of ye present service, as the consequence it may be to his Ma^{ty} that all occasion of repineing be removed from such as are employed'.[5] Oppenheim found 'no trace of the sale of places during the Commonwealth, but the custom was reintroduced with the other fashions of the Restoration'.[6]

While these themes may be traced back to Charles I's personal rule in the 1630s, this itself stemmed from earlier reigns. Tudor and Elizabethan legacy was embedded within 1630s naval administration; Charles's personal rule without parliament, just three decades after Elizabeth I, was not such a remarkable policy.

This article will analyse top-down naval administration under James and Charles, as discussed by Oppenheim, Tanner, Aylmer, McGowan, and Rodger, the effects of the 1620s naval campaigns upon financial planning, and detailed Navy Board correspondence to the Admiralty concerning dockyards, ships and chips. The chief primary source is Navy Reports 1618-1637 at the British Library.[7]

Abuses

That the reign of James I (1603-25) 'was perhaps the worst and most corrupt in the history of naval administration' was established by the 1608 and 1618 Commissions of Enquiry.[8] The most spectacular abuse identified in 1608 was the building and commercial venture to Spain of *Resistance* in 1605, undertaken by navy treasurer Sir Robert Mansell, surveyor Sir John Trevor and master shipwright Phineas Pett. However, vague clarification of their public and private responsibilities allowed James to avoid dismissing them. Their placemen allowed them to collect fees throughout naval administration.[9]

McGowan warned in 1971 that 'one must be wary of applying twentieth-century moral standards to the actions of officials 350 years earlier'. It was customary to accept a gift (not regarded as a bribe) made *after* a post was awarded, as a sign of gratitude; New Year gifts were common; fees were routinely paid for entry to a post, completion of documentation and passing of accounts.[10] Bribery *per se* was distinguished by Aylmer as 'an attempt to persuade an official to follow a course of action other than that which he knew he ought to have followed'.[11] If proven this could supply the evidence needed for removal from office and disgrace, such as Lionel Cranfield in 1624 and attempted against Buckingham in 1626. Pepys's own rationale was similar, approving

Fig. 2. Marble sculpture of Charles I by Honor Pelle, 1684. Ann Coats, 2009.

Table 1
Admiralty and Navy Annual Salaries, Pensions and Perquisites in the 1630s[19]

Officer	Fee or Salary £	Pension, Annuity, other office £	Gratuities & fees £	Annual Total £
Admiralty Secretary Edward Nicholas	50	200	'unknown but heavy'	1,500 est. min.
Navy Secretary Sir Albertus Moreton in 1625	700 pa for pleasure for Intelligence	Granted other expenses as Secretary; 500 pension for life		c. 2,000 net
Treasurer	221 or 271 including allowances	200-653 poundage early 1630s; possibly 900-1,000 1637-40	Gratuities at least 350	1,600 in 1640
Comptroller	155 (138 net)	Av. 35 expenses; 120 from 1639	Not known	At least 258 late 1630s
Surveyor	145 (128 net)	Av. 35 expenses	Not known 100 from 1639	At least 228
Clerk	100 (92 net)	Av. 32 expenses; 80 from 1639	Not known	At least 172
Lord Treasurer's Remembrancer	62	98; 3 livery servants	Not known	At least 160
Prest Auditor	67	5 livery servants	c. 180 minimum 1638-9	At least 300

the 1618 report and aware of the puritan ethos during the interregnum.[12] In 1644 at Deptford a lecture on 'saving truths' was given every Wednesday morning to the workforce as part of their work time, and it was proposed: 'That prophane swearing, cursing, drunkene—s and other misdemeanours which are frequently practised by many imployed in the yard bee severely punished'.[13] Aylmer calculated that naval administration fees in the 1630s amounted to c.£1,000 a year.[14] As fee exactions caused dissatisfaction, royal commissions on fees 1610-11, 1623-4 and 1627-40 demonstrated both the desire and the substance of reform.

Appointments and promotions were made by the king, department head or subordinate officials. Occasionally they were made on merit. Aylmer distinguished three influences upon appointment to office: patrimony (extended family connections), patronage (friendship, common interests or 'that attachment by a lesser man to the following of a greater') and purchase (citing Trevor-Roper's rate of 3.5 years of the office salary as a norm, but varying according to circumstances). Purchase often accompanied one of the other two and it was safer to have more than one patron. Aylmer maintained:

Appointment by patronage and other personal contacts was not incompatible with men rising on their merits....the methods of entry were more haphazard than those with which we are familiar today; it does not follow that they invariably produced worst results.[15]

Proprietorship of public office complicated post-holding, promotion and custodianship of papers, epitomised at the end of Pepys's own career as Admiralty secretary. By 12 July 1689 he had returned specified papers to Phineas Bowles, Secretary to the new Admiralty Commission, but retained many public papers.[16]

Some offices were hereditary, such as Tellers of the Receipt.[17] Table 1 shows that more income derived from fees and perquisites than salaries, a motivation for personnel to purchase such offices through patrons, and was believed by contemporaries to distort performance from the principle of duty when in office. From 1616 to 1628 most naval patronage was controlled by George Villiers, Duke of Buckingham (1592-1628), who, as Lord Admiral received Admiralty prize money after 1619, yielding an estimated annual £3-4,000 in wartime.[18]

During Charles's reign the navy, ordnance and wardrobe administrations had the 'worst contemporary reputation for waste and fraud'.[20] Pluralism and selling of offices were included in the Commons' attempted impeachment charges against Buckingham in 1626. John Pym (receiver-general for life - a recipient of fees), summed up the arguments succinctly: the 'wrong type of men will buy entry, and then buy promotion, their wealth making up for their lack of merit'; 'ambitious men' will spend their time raising money to advance themselves rather than to the profession itself; 'the richest men will always prevail over the most deserving'; those who buy office will aim to 'maximise their annual value and to sell again for more than they gave'; those appointed on merit will be dismissed if another is prepared to pay.[21] As king's servants, officials were expected to be loyal to their sovereign but they also advanced their personal interests and political power.[22]

Aylmer concluded that there was less evidence of selling offices (by the king, patronage-holder or incumbent) in the 1630s compared with 1615-23, although this was modified by royal financial needs.[23] John Coke, deputy treasurer of the navy 1599-1604, considered the Lord Admiral's monopoly of patronage the main cause of 'disorders in the navy'; in November 1618 he urged Buckingham to prohibit the sale of naval offices and reversions to control patronage for 'the reputation and flourishing estate of the navy'.[24] In 1627 Charles set up a Commission on 'Exacted Fees and Innovated Offices' to address grievances against exactions and raise money by fining offenders. Taking fee levels of 1568-69 as a standard, it characterised later or higher fees as exactions and new offices as innovations. An executive committee was appointed in December 1627 to take evidence under oath.[25] Around 1629 administrative reforms were begun in the Household.[26] In April 1638 Charles made Algernon Percy, Earl of Northumberland, Lord Admiral until his son James reached the age of 22. If James had died the king could have re-disposed of the office.[27] Admiralty commissioners declared against reversions in March 1637: from 1638 no naval administration appointments were to be made for life, but only during the king's pleasure. The trend under Charles I was therefore for shorter and less secure tenure.[28]

Loyalty in royal service derived from concepts of honour, attached to aristocratic and gentlemanly status. Aylmer concludes: 'Most early Stuart administrators, especially those in the more senior positions, were born of armigerous families, or at least of those on the way to becoming armigerous'. Perceptions of status in this period were fluid, affected by James I's creation of baronetcies in 1611, his 'profuse granting of titles' and Charles I's imposition of fines on those with annual incomes of £40 who had not been knighted at his coronation in 1626. The title of 'Gentleman' was restricted officially to younger sons and brothers of esquires (descended from armigerous knights, holders of state offices, MDs, DDs and MAs), but in practice was assumed by 'anyone who felt himself to be a member of the gentry or wanted to pass as one'. He asserted that merchant magnates called themselves knights, esquires or gentlemen. Of his sample, 55.1% had been to university and 47.5% had attended Inns of Court.[29] This issue caused problems of precedence which overlaid other hierarchies, demonstrated by the internal Navy Board quarrel for precedence in 1628 because treasurer Russell was a junior baronet while controller Slingsby a lesser knight bachelor.[30]

Table 2
Commission for the Survey of the Navy 1618-1623[31]

Sir Thomas Smythe	City merchant and a governor of Muscovy and East India Companies. His name headed the patent roll.
Sir Lionel Cranfield	Master of the Household and Wardrobe, spokesman and chief commissioner, Lord Treasurer 1621
Sir Richard Weston	A former collector of 'little customs' in London, later Lord Treasurer, 1st Lord Weston, 1st Earl of Portland 1633,[32] Lord Treasurer.
Sir John Wolstenholme	A London customs farmer, member of the Company of Merchant Adventurers
Nicholas Fortescue	A commissioner of James I's household, later Chamberlain of the Exchequer
John Osborne	A Lord Treasurer's Remembrancer[33]
Francis Gofton[34]	A Lord Treasurer's Remembrancer, an Auditor of the Prests[35] and Navy accounts
Richard Sutton	An Auditor of the Prests and Navy accounts
William Pitt	City merchant, member of Muscovy Company, a former officer in the Exchequer of Receipt
John Coke	Deputy Treasurer of the Navy 1597-1604 under Sir Fulke Greville
Thomas Norreys	Master Shipwright formerly employed in the navy
William Burrell	East India Company Master Shipwright for 7 years

According to McGowan, the 1618 Commission 'was an almost unqualified success'.[36] Collectively the commissioners were experienced in royal household management, commerce, finance or maritime affairs. Its patent was in Sir Thomas Smythe's name but Sir Lionel Cranfield, whose first patron was Charles, 1st Earl of Nottingham, acted as chief commissioner until 1623.[37] According to Young, it was 'basically a subcommission of the one Cranfield had headed in the reform of the Household'.[38] Despite Cranfield's 'overshadowing presence', Young showed that Coke, out of office from 1604 until appointed to the commission in June 1618, 'dominated the work of the naval commission', as 'Cranfield had neither the time nor the experience necessary to direct the reform of the navy'.[39] Like his patron Fulke Greville, navy treasurer, Chancellor of the Exchequer, and Treasury commissioner under James I, Coke believed in the established church to give political stability and was called a 'Puritan' because of his rigid moral principles. Politically he believed in hierarchical subordination of all officers to the king, who could change or dissolve jurisdictions 'for public good'.[40] Religious persuasion operating within naval patronage was clearly significant in the 1630s, and Aylmer suggested that religious motives 'seem sometimes, but not always or necessarily, to have been complementary to material ones'. In his assessment, the

> *appointment of Earl of Northumberland as Admiral of the Fleet in 1636-7 and Lord Admiral in 1638 meant that by 1640 the co-Treasurer, Surveyor and patronage Secretary were all Puritans of one kind or another, while the Clerk of the Acts may have had leaning the same way.*[41]

Pluralism was common, leading to the appointment of deputies. Sir Thomas Aylesbury, Surveyor of the Navy 1628-32 and co-Master of the Mint, employed a 'man of business'.[42] Edward Nicholas was clerk of the Council extraordinary 1626-35 and ordinary 1635-41, the Lord Admiral's secretary 1625-40 and secretary to the Fishing Association c.1630-40; to Aylmer a 'good example of a conscientious administrative careerist'.[43] Sir William Russell was treasurer of the navy 1618-27 and 1630-42; also a leading merchant of several companies, a customs farmer and collector of duties on silk and imported fabrics.[44]

The 1618 report, drafted by Coke, but presented by Cranfield to the Privy Council, James I and Buckingham on 29 September, pinpointed causes for waste in bad management: 'Great works are taken in hand, and multitudes kept in pay when neither materials nor moneys are provided, and this made the wages of shipwrights and caulker grow to…£17,372 2s 5d' in 1614-1617.[45] For this money eight new ships might have been built, but instead ships had decayed. Stores had decayed due to insufficient inspection on delivery, weights at dockyards were inaccurate, leading to deficient received stores, and accounts recorded more stores entering than had actually been received. Boatswains' low wages encouraged them to steal cordage. The building and beautifying of officers' houses and selling of offices wasted funds and misused stores. The 'officers are raised above their own orb and the clerks and inferiors come into their rooms, and their places again are subdivided and multiplied into those many new offices specified in our books'.[46] Extra and duplicated officers' wages, expenses, fees and allowances wasted money. It attributed

> *the chief and inward causes of all disorders to be the multitude of officers and poverty of wages, and that the chief officers commit all the trust and business to their inferiors and clerks, whereof some have part of their maintenance from the merchants that deliver in the provision that they are trusted to receive.*[47]

The commission surveyed stores, moorings, dockyards, ships and accounts, and took depositions on oath to ascertain abuses. Since 1613 only annual abstracts of expenditure existed and no quarterly dockyard books had been made up since 1615.[48]

The report resulted in changes at the top. On 10 May 1618 Sir Robert Mansell sold his office of treasurer, held since 26 April 1604, to Sir William Russell.[49] The report avoided criticising Nottingham as Lord Admiral and proposed a scheme to reduce annual naval expenditure from £53,000 to £30,000, with no reduction in naval capacity.[50] At the beginning of October 1618 Buckingham and Nottingham, Lord Admiral since 1585, were joint Lord Admirals, but at the end Nottingham agreed to retire with an annual royal pension of £1,000 and a payment or gift of £3,000 from Buckingham. George Villiers, Duke of Buckingham was a courtier who, according to McGowan, 'had indicated an interest in the reform of the navy and the post of Lord Admiral'. His Instructions expressed the need for 'administrative continuity' and 'a Gen$^{l.}$ System of the Sea Service in all its Parts'. They were prompted 'cheifely from the want of good Ordinances & Instruc–ions to direct the Service'.[51] Coke could persuade Buckingham to promote reform by appointing naval commissioners to replace the tarnished principal officers, who by 'purchasing their places at dear rates, must strain to cover the same, or live by the loss'.[52]

Judgement on the report was given in the Privy Council on 2 November 1618, before James, Prince Charles, Buckingham, Secretary of State Sir Robert Naunton, Fulke Greville and the principal officers. They, comptroller Sir Guildford Slingsby and surveyor Sir Richard Bingley, in post since 7 May 1611, had declined to implement the report, and the clerk of the navy, Sir Peter Buck (and/or John Legatt since 1604), in office since 1596, were all implicated in its findings and became scapegoats, sequestered from office. They retained office through their patents and remained on full pay but could not perform any functions.[53]

The report recommended raising the wages of masters, boatswains, gunners, clerks of the cheque and officers' clerks to improve honesty and accountability, recruit more able men and save money.[54] This point was made by Hollond in 1638:

> *How is it possible for a boatswain, having*

a wife and three, four, or five children depending upon his labour, to maintain himself and them with 20l. per annum wages, without clenching, changing, selling, wasting, and purloining of his Majesty's cordage and other stores committed to his trust? [55]

It found overwhelming evidence of the sale of offices, for instance Hugh Lidyard was made Woolwich clerk of the checque by Sir John Trevor, Navy Surveyor 1598-1611, paying him £20 a year and a hogshead of wine. Another witness deposed that few 'come freely to their places'.[56] Proof was found of dead ships: £17,000 had been paid for a new ship to replace *Bonaventure* and £63 paid annually for maintenance but no such ship was built; *Advantage* was charged £104 9s 5d five years after being burned; *Charles* was charged £60 16s 10d two years after being disposed of.[57]

The report distinguished the 'ancient institution', where all officers had 'special duties', and were 'few and yet all was kept in order by their daily attendance and continual accounts', from 'recent innovations' whereby they had only general duties, used deputies, and multiplied offices and expenses. It called for 'reducement to the ancient frame' of Henry VIII.[58]

The commission urged that such reform could not be achieved through the principal officers, but only 'by enlarging the Commission that already is on foot, giving power to the Commissioners to put into execution their propositions for restoring the navy'.[59] Fundamental points were the elimination of dead ships and dead pays, instigation of appropriate salaries for officials and a rebuilding programme.[60] To illustrate the top-down, bottom-up dynamic, Coke raised these points first with Buckingham on 17 October 1618. He forwarded them, endorsed by James, to Secretary of State Naunton, who sent them to the Privy Council: 'that they put into practice these directions of his Majesty'. Coke then received his orders from the Council. Later he drafted Fulke Greville's letter to Buckingham reporting what had been done. As Young says, 'orders had to proceed downward from the top of the court hierarchy'; but 'Coke, who alone knew what orders had to be given, was not at the top but the bottom.' Young concludes that 'the crucial link in the circle was Buckingham', who 'made the whole process possible by virtue of his influence over James, his willingness to accept Coke's advice, and his desire to appear knowledgeable and forward in reform of the navy'.[61] The Greville/Coke letter of late October urged that 'the officers of the navy cannot be made fit instruments to put that in practice which discovereth their shame'.[62]

The report proposed a fleet of 30 ships costing £30,000 a year for five years (compared with the cost of navy in 1618 of £53,000), then £20,000 annually.[63] It promised:

- *This navy will contain at least 3050 tons more than the navy of Queen Elizabeth when it was greatest and flourished most.*
- *The navy as now it is decayed may in five years be raised to this perfect establishment, and in that short time of restoring, not cost His Majesty so much as was spent in the time of decaying and afterwards be maintained in harbour and at sea with a far less charge.*
- *The dockyards may no more appear to strangers or others as wrecks, or as empty or ruined houses but in their complete equipage ready prepared to set out against the enemy, upon every alarm or command from the state.*[64]

From the beginning of December 1618 the Commissioners assumed responsibility for the next five years at no charge to the king. Sir William Russell, in office since May 1618, was retained as treasurer under their control. Buckingham's patent as Lord High Admiral was issued on 28 January 1619 and on 12 February 1619 the Commission became a board of governors, with Coke effectively the principal administrator.[65]

The programme produced two new ships in November 1619, *Reformation* and *Buckingham's Entrance*, launched at Deptford by James. Two more ships were launched in 1620 and a new double dock was begun in 1618 at Chatham, 330 foot long and costing c.£4,000.[67] The objectives of building two ships a year and running the ordinary on budget were met admirably until undermined

Table 3
Ships built by the Navy Commissioners 1618-28[66]

Antelope 1581 rebuild	34 guns	1618		*St Anne*	guns	1626
Happy Entrance	30	1619		*Fortune*		1627
Constant Reformation	40	1619		*First Whelp*	14	1628
Victory	40	1620		*Second Whelp*	14	1628
Garland	34	1620		*Third Whelp*	14	1628
Swiftsure	44	1621		*Fourth Whelp*	14	1628
Bonaventure	32	1621		*Fifth Whelp*	14	1628
St Andrew	42	1622		*Sixth Whelp*	14	1628
St George	42	1622		*Seventh Whelp*	14	1628
Triumph	44	1623		*Eighth Whelp*	14	1628
Mary Rose	26	1623		*Ninth Whelp*	14	1628
Charles	14	1623		*Tenth Whelp*	14	1628

by inadequately funded wars and campaigns and a shortage of skilled naval personnel.[68] However, Hollond criticised work done by contract, or the 'great', rather than day work, noting of the *Whelps* that 'some whereof have miscarried, others forced, not a year after their building, to be brought into dry dock to be strengthened under water, &c.'. Tanner and Oppenheim noted that they were built by nine master shipwrights at £3 a ton, 'described as built in haste "of mean sappy timber", and only one of them lasted until the time of the Commonwealth'.[69]

In 1619-21 an unsuccessful attack on Algiers cost over £30,000, taking £3,000 out of the 1622 ordinary budget and James I declared war on Spain 10 March 1624.[70] The Anglo-Dutch fleet, hampered by lack of money, unskilled personnel and insufficient provisions, blockaded Cadiz ineffectually in October and November 1625.[71] To raise money Captain John Pennington cruised the Channel in February 1626, capturing ships and cargoes worth £50,000. Lord Willoughby captured French prizes in September, but failed to capture a Spanish *flota*, while the English wine convoy was captured. The City of London raised a defensive squadron of twenty 'ill conditioned' ships in October, with which Pennington attacked Le Havre.[72] Between 1626 and 1630 privateering captured up to 1,000 foreign prizes, but the English south-west and Irish coasts and shipping were raided by French, Spanish and Dunkirk privateers and Barbary corsairs from Algiers and Tunis, with the navy largely ineffectual as protection.[73]

In 1623 the commissioners reported their progress after five years: in 1618 they had found 23 and ten unserviceable ships, four decayed galleys and four hoys; they now had 35 serviceable vessels, plus the galleys and hoys, costing £30,000 a year, including building ten new ships. They aimed to maintain 30 sea-going ships.[74] Master shipwrights William Burrell built new ships at Deptford, Phineas Pett at Chatham. At Chatham two mast docks were built in 1619 and 1620, each 120 ft long and 60 ft wide, with six acres of ground. A further extension was used to build the doubledock, costing £2,342, a ropehouse and brick and lime kilns. In 1623 a single dock was built.[75] The Navy Commissioners were continued until 1628.[76]

Charles I

Charles I, James's second son who became heir in 1612 on the death of the favoured Prince Henry, succeeded James I on 27 January 1625. Charles's abortive 1623 expedition to Spain with Buckingham cost £20,000, bringing his bride Henrietta Maria to England in 1625 nearly £36,000, and the Cadiz campaign in 1625 £500,000. The new régime was disliked because of Buckingham's perceived venality and the unpopularity of their Spanish expedition. Parliament voted £300,000 in 1624, and but only two subsidies amounting to £140,000 in 1625, and no customary Tunnage Poundage for Charles's lifetime.[77] In July 1625 Coke conveyed the previous year's military expenses (£280,000) to parliament and what was required to clear debts (£573,000). Parliament did not grant the funds and recessed because of plague. Lack of money delayed sailing to Cadiz until

8 October 1625, and the quality of leadership, ships, seamen, provisions, ordnance and clothing was disastrous. A hundred ships, mostly hired merchantmen, were commanded, mostly ineffectually, by Sir Edward Cecil, Viscount Wimbledon. The plan to blockade Cadiz was inept: the attack was disordered, Spanish merchant ships escaped and the merchantmen refused to fight. It was estimated at £300,000 but cost half a million.[78]

In 1626 Coke asked parliament for over £1m to meet war expenses and debts, but a Commons antagonistic to Buckingham offered only a quarter of this sum, conditional upon Charles meeting their grievances and tried to impeach Buckingham for selling offices.[79] Charles responded by dissolving parliament. With Willoughby's 1626 expedition costing £90,000, Charles ordered a forced loan in 1627 and Russell negotiated a loan of £95,000 in 1628, but Charles's third parliament of 1628 forced him to accept the Petition of Right which declared forced loans illegal and in 1629 refused Tunnage and Poundage. In March Charles dissolved parliament and called no more in his reign.[80]

Failure to pay seamen's wages exacerbated the navy debt: £4,000 a month accrued because £14,000 was needed to pay off the campaign fleets. It led to mutiny in Portsmouth in 1626. Seamen's monthly pay was raised from 10s to 15s (with deductions for the Chatham Chest, a preacher and surgeon), but the victuallers had not been paid and seamen with pay tickets and Chatham shipwrights, who had not been paid for a year, besieged the navy commissioners for 20 days. In October 1626 an order in council declared that crews should be given a 'competent reward', but none was forthcoming; in October 1642 Parliament announced that one third of the prize value should be divided between the officers and crew, but much of this was deducted by Admiralty officers.[81] Another commission was set up in December 1626 to examine the state of the navy, specifically the new ships built by Burrell. The report found failings, but five continued to the restoration.[82]

In 1627 Buckingham sent a fleet to protect French Huguenots in La Rochelle and watch out for a Spanish Armada, its costs of £70,000 raised privately by Buckingham and Russell. The Huguenots at first refused his help, so from July to October 1627 he besieged Fort St Martin on Île de Ré which guarded the approach to La Rochelle, but failed to capture it or prevent the French from supplying it because of poor supply lines. He withdrew, returning with only 2,989 men out of 7,833 embarked.[83] He prepared two more fleets to relieve La Rochelle in 1628, but could not attract seamen, despite the increased wages, because of unpaid wages, unfit food and sickness, and because merchant shipowners were paying 30s a month. Parliamentary attacks caused further delays. The waiting fleet in Portsmouth became demoralised, leading to mutiny over pay, clothing and victuals, and Buckingham's assassination on 23 August 1628 by Lieutenant John Felton.[84] La Rochelle surrendered on 18 October.[85] The Treaty of Madrid ended war with Spain on 5 November 1630.

After Buckingham's death the Admiralty was put into commission. A board comprising the Lord Treasurer, the Chancellor of the Exchequer and two Secretaries of State directed the Navy Board. According to Oppenheim it was, 'in reality a committee of the Privy Council' and 'even more reliant on the capacity of the Principal Officers' than their predecessors, but also distrustful. They met at Wallingford House until 1634 or in the Whitehall Council Chamber.[86]

Fig. 3. Île de Ré and a plan of Fort St Martin in the approaches to La Rochelle. Ann Coats, 2008.

Table 4
Lord Admirals (ex-officio Privy Councillors) Commissioners and Admiralty Secretaries 1585-1649[87]

	Post	Appointed	Notes
Charles Howard	Lord Admiral	8 Jul 1585	Lord Howard of Effingham, 1st Earl of Nottingham until October 1618
George Villiers	Lord Admiral	28 Jan 1619	1st Marquis of Buckingham, 1st Duke of Buckingham 1623, died 23 Aug 1628
Richard Weston	Lord of the Admiralty	20 Sept 1628	1st Lord Weston, 1st Earl of Portland 1633, Lord Treasurer, died March 1635
Robert Bertie	Lord of the Admiralty	10 Apr 1635	1st Earl of Lindsey, Lord Chamberlain
William, Earl of Pembroke	Lord of the Admiralty	20 Sept 1628	Until 1632. Lord Steward.
Edward Earl of Dorset	Lord of the Admiralty	20 Sept 1628	Lord Chamberlain to the Queen
Algernon Percy	Acting Lord Admiral	13 Apr 1638	4th Earl of Northumberland. 17-year tenure until James Duke of York became 22. He did not purchase the office. His appointment may have resulted from his complaints in 1636, leading to a commission. His commission was revoked by Charles I 25 June 1642
Francis Cottington	Lord Admiral	20 Nov 1632 1635	Weston's protégé in the 1630s. Chancellor of the Exchequer 1634, Master of the Wards 1st Lord Cottington appointed by Charles I, but had no access to the Fleet.
Sir Henry Vane	Lord Admiral	20 Nov 1632	Diplomat, member of Privy Council since 1630. Succeeded Sir John Coke as Secretary in 1640.
William Juxon	Lord of the Admiralty	16 Mar 1636	Bishop of London, Lord Treasurer; he did not purchase the office
Algernon Percy	Senior Admiralty Commissioner	1638	4th Earl of Northumberland; removed by the king in 1642
Robert Rich	Lord High Admiral	July 1642-45	2nd Earl Warwick, puritan, brother of Henry Rich, 1st Earl of Holland. Resigned April 1645.
Algernon Percy	Senior member Parliamentary Admiralty Committee	19 Apr 1645	4th Earl of Northumberland (Warwick the effective chairman)
Robert Rich	Lord High Admiral	29 May 1648	2nd Earl Warwick
Council of State	Admiralty & Navy	23 Feb 1649	Lord Admiral's powers transferred
Sir Edward Conway	Admiralty Secretary	1623-28	1st Lord Conway, promoted to the Lord Presidency of the Privy Council
Sir Albertus Moreton	Admiralty Secretary	1625	Previous overseas experience, died 1625
Sir John Coke	Admiralty Secretary	1625	Knighted September 1624; 'virtually dismissed' in January 1640
Dudley Carleton	Admiralty Secretary	20 Sept 1628	Viscount Dorchester, Vice-Chamberlain of the Household, previously a foreign agent and ambassador, died Feb 1632
Sir Francis Windebank	Admiralty Secretary	20 Nov 1632	Weston's protégé, succeeded Carleton, knighted 1632.
Edward Nicholas	Admiralty Secretary	1625	1638
Thomas Smith	Admiralty Secretary	15 Sept 1642	Northumberland's Secretary

Fig. 4. The Duke of Buckingham's house in Portsmouth High Street, where he was assassinated in 1628. Ann Coats, 2012.

Charles and Imagery

Charles's attitude to kingship and authority was expressed aesthetically by his lavish art collection, inspired by the collection acquired by his brother Henry before he died, and furthered by Buckingham. He invited Orazio Gentileschi as court painter in October 1626 and appointed Van Dyck principal painter in 1632. Charles spent £18,280 12s 8d buying a renaissance art collection from Mantua in the late 1620s through diplomats based in Venice and ruined financier Philip Burlamachi, already overstretched by the Ré campaign. The most

Fig. 5. Memorial of the Duke of Buckingham in St Thomas's Church (Cathedral), Portsmouth. Ann Coats, 2012.

iconic images of his reign were Van Dyck's *Charles I with M. de St Antoine* (1633) and Hubert le Sueur's equestrian statue of Charles I.[88]

Charles's projected his power aspirations through his navy. The Channel squadron was instructed to impose salutes to the English flag and enforce Dutch shipping licences in waters claimed as sovereign by the English crown, recalling medieval attempts to enforce jurisdiction over the 'Narrow Seas'. His greatest ship, *Sovereign of the Seas*, epitomised his aspirations of grandeur.[89]

During Charles's reign, in Aylmer's view, the 'Navy was a bigger and more important department' of state than the Mint, Exchequer, Ordnance or Armoury. It was directed by the King and Privy Council through the Lord Admiral. Charles became more involved in naval affairs after Buckingham's death. State papers show that he 'exercised a constant personal supervision in naval affairs, sometimes overruling the opinions of his officials in technical details of which he can have possessed no special knowledge'.[90]

Navy Board Restored 1628

The Navy Board was established as in 1546 as a household office of Henry VIII from officers existing since 1514 as

> the lord high admiral's council of advice..."to whom the lord admiral properly directs all his commands for his Majesty's service, and from whom it descends to all other inferior officers and ministers under them whatsoever."...In practice they enjoyed very large administrative powers, for they were authorised "to cause all ordinary businesses to be done according to the ancient and allowed practice of the office, and extrordinary according to the warrants and directions from the lord admiral and the state", but in theory they existed only to carry out the admiral's instructions[93]

Also known as the principal officers, they consisted of the treasurer, who held the highest status because he controlled the money, the comptroller, nominally senior to the others: the surveyor, responsible for materials, the clerk of the ships or navy, the

Table 5
Navy Board (Principal Officers) restored 1628[91]

	Post	Appointed	Notes
Sir Sackville Crowe	Treasurer (accounted to the Prest Auditors)	5 Apr 1627	Removed 1629
Sir William Russell	Treasurer 1634	21 Jan 1630 Signed from 29 Nov 1627	Bought the post from Sir Robert Mansell 10 May 1618; held until 5 Apr 1627. Until 12 Jan 1639 then jointly with Sir Henry Vane Jr until Dec 1641. Also a member of the Muscovy Company
Sir Henry Vane Jr	Treasurer	12 Jan 1639	Jointly with Sir William Russell until Dec 1641, then alone until 5 Aug 1642. Appointed Secretary of State in 1640, for which he did not pay.
Sir Gylford Slingsby	Controller (checked the Treasurer and other officers)	Feb 1628	Died 1632
Sir Henry Palmer Jr	Controller	Signed from 13 Apr 1632	11 Dec 1639, then jointly with Sir George Carteret until Nov 1641
Sir Thomas Aylesbury	Surveyor	Feb 1628	Claimed precedence from his family status. Surrendered / bought out 19 Dec 1632.
Kenrick Edisbury	Surveyor	19 Dec 1632	Granted 17 Dec 1632; died 26 Aug 1638
William Batten	Surveyor	26 Sep 1638	Appointed 'during pleasure', not for life. Grants 20 Nov 1639, 20 Jun 1660; until Sep 1642, then continued as Navy Commissioner Sep 1642-1648, often at sea; knighted 28 May 1648; died 5 Oct 1667
Sir Sampson Darrell	Victualler (contracted, salaried and holding a patent; not under the Treasurer)	1630	Had been joint Surveyor of Victualling with Sir Allen Apsley since 1623. Died 1635 owed money by the crown. His claims of £69,436 in 1626 and £94,985 in 1627 were rejected.
John Crane	Victualler	20 Nov 1635	A royal household official; wished to terminate his contract in Mar 1638; was heavily in debt by 1640, the crown owing him £3,825
William Burrell	Assistant Extra Principal Officer	Signed from 29 Nov 1627	Previously master shipwright to the East India Company, died 1630
Phineas Pett	Assistant Extra Principal Officer	1629	Son of Peter (d. 1589). 1621 joint master shipwright at Chatham; died Aug 1647
Sir Kenelme Digby	Extra Principal Officer	Oct 1630	Coke's representative; 'no defined duties on the Board' (Clowes)
Phineas Pett	Extra Principal Officer	Jan 1631	
Peter Buck	Clerk of the Navy (junior Principal Officer)	10 Jul 1596	Granted for life; the patent salary was £33 6s 8d. Knighted 1604. Alderman and Mayor of Rochester. Died by 21 March 1625.
John Legatt	Clerk of the Navy	17 Apr 1604	Granted in reversion after Peter Buck; did not succeed
Dennis Flemyng	Clerk of the Ships	21 Mar 1625 Signed from 29 Nov 1627	Surrendered 11 Feb 1639
Thomas Barlow[92]	Clerk of the Ships Exchequer fee £33 6s 8d	16 Feb 1639	Granted for life; revoked 13 Jul 1660
	2 Assistant Clerks		
John Wells	2 Principal Storekeepers		Complained that his lodgings at the Navy Office had been seized by Slingsby

oldest but least powerful officer, and the surveyor of victualling.⁹⁴ According to Oppenheim the duties of the principal officers in 1617 were:⁹⁵

- *Treasurer: financial and general superintendence*
- *Comptroller to check the accounts of the treasurer and surveyor; inspect stores and storekeepers' book*
- *Surveyor to inspect ship, wharves, houses, chain and ships on return from the sea; draw out indents for ships' stores*
- *Clerk to keep minutes of resolutions and attend the yearly general survey*

The board was restored in February 1628, although correspondence shows that Russell, Burrell and Flemyng were signing letters on 29 November 1627.⁹⁶ They met first in St Martin's Lane, north of Charing Cross, but in 1630 rented rooms in Mincing Lane at £36 a year which cost £150 to furnish.⁹⁷ This distance of 2.5 miles from to the City of London, where merchants and shipowners were based, to the Admiralty at the heart of the Court became a tangible barrier to communication. It continued to separate the Admiralty and Navy Boards spatially until 1789, when the Navy Board moved to Somerset House.

Retired Admiral Sir William Monson's ideal was a treasurer who was an experienced merchant/shipowner who could raise money on his own credit; a first class master shipwright and sea captain for the two surveyors; and a reliable and experienced clerk as joint comptroller/clerk.⁹⁸ Sir Sackville Crowe, Treasurer 1627-29, took poundage but did nothing and took £3,000 from the Chatham Chest. In 1630 the treasurer was granted poundage of 3d on all payments made by him, including wages (previously just for merchants' bills) and a house at Deptford. In 1634 his exchequer fee was increased from £270 13s 4d to £645 13s 4d.⁹⁹ The salaries of the other three officers' had not been increased since their foundation on 24 April 1546: the comptroller £50 a year plus travelling expenses and two clerks, the surveyor £40 plus expenses and one clerk, and clerk of the Ships, £33 6s 8d plus travelling expenses. They also had perquisites, fees charged for transactions.¹⁰⁰

The Admiralty Lords considered that the principal officers' sympathies lay 'rather with their subordinates than with the King's interests'. It was natural for them to feel affinity with their subordinates, as they had risen from this background and worked with them on a daily basis; their correspondence is full of empathy, whereas the lords were courtiers who only visited dockyards on ceremonial occasions. Oppenheim considers that 'the conditions under which they worked were not favourable to success in management': their books were kept in their own residences, there was insufficient time to keep the books, and they had to buy some stores from those who held royal patents.¹⁰¹

Fig. 6. Le Sueur's equestrian statue of Charles I, 1633. Ann Coats, 2010.

Rodger considers that despite 'a quarrel over precedence' between the surveyor (Aylesbury), the treasurer (Russell) and the comptroller (Slingsby), 'a number of able and reasonably honest men' including Edisbury, Digby, Russell, Darrell and Crane were appointed.[102] Hollond, Edisbury's protégé, reported that in 1634 Palmer, Pett and Fleming 'were suspended for selling government stores', offences similar to those they were supposed to be preventing, but they were punished leniently.[103]

Despite officials' wages in arrears, embezzlement and shipkeepers being too old, young or ill-trained, by 1631 Charles could see improvements.[104] Chatham Dockyard was described as:

admirably organised, so that all the requisite apparatus is always in readiness, carefully guarded and deposited...divided into several compartments, each one containing everything necessary for arming a ship. The arms or device of the name of each ship is placed on the door of these apartements, and thus distinguishes what belongs to them.[105]

Hollond's 1638 *Discourse* recorded both improvements and continuing evidence of abuses. Pursers' offices were the most eagerly sought because of the amount of money which could be made in post, particularly by selling beer, stores and clothing.[106] On the positive side Hollond cited the 'vigilance and fidelity' of a clerk of Chatham ropehouse who through his inspection of new parcels of hemp, reduced the vendor's payment.[107] The Duke of Buckingham's Instructions made this a duty of dockyard officers.[108]

In Hollond's opinion the 'memory of their first institution' (1546), had become 'blurred' and distinct duties had been lost. He questioned how one principal officer living in London with a general duty could combine particular duties in the yards at Portsmouth, Deptford, Chatham and Woolwich (to attend pays, survey and certify stores) 'at one and the same time'.[109] He proposed an assistant to the principal officers at each dockyard, to save the principal officers time, travel and expences, citing Captain Joshua Downing, his early patron, as a positive example of an assistant principal officer.[110] During the interregnum and restoration resident commissioners were established at the dockyards.

Ship Money

Lacking parliamentary funds, Charles used ship money as an extra-parliamentary tax levied by writ to raise money for the peacetime navy. Three ship money writs were issued 1634-6. The first were issued to London and the ports on 20 October 1634. Charles extended the second writs to inland towns on 4 August 1635 and the third on 9 October 1636, enforced by sheriffs and justices of the peace. 95% of the 1635 assessments was collected in 1636. During 1635-40 ship money raised an annual average of $c.$£107,000, 'far in excess of any previous direct tax in peacetime'. In 1634-40 it totalled £800,000, £200,000 more than Charles's parliamentary taxation. However, it could only be collected through the 'voluntary co-operation of a

Table 6
Navy Board Ships built 1629-40[113]

Vanguard 1615 rebuild	40 guns	1631	*Roebuck*	10 guns	1636
Charles	44	1633	*Greyhound*	12	1636
Henrietta Maria	42	1633	*Providence*	30	1637
James	48	1634	*Expedition*	30	1637
Unicorn	46	1634	*Sovereign of the Seas*	100	1637
Swallow	34	1634	*Prince Royal* 1610 rebuild	64	1639-41
Leopard	34	1635	*[Red] Lion* 1609 rebuild	42	1640

hierarchy of part-time, unpaid officials: lord and deputy lieutenants, sheriffs, justices of the peace, high and petty constables, overseers of the poor and churchwardens'. When this cooperation declined, only 80% was collected in 1639 and 21% in 1640.[111]

The tax paid for summer campaigns: in 1635 Lord Lindsey commanded 19 navy ships and five merchantmen; in 1636 the Earl of Northumberland commanded 27 navy and three merchant ships in the Channel and North Sea; it furnished the two pinnaces used by Captain William Rainsborough to blockade Sallee (Rabat in Morocco) in 1637 and free 340 English captives. Shipbuilding, dockyards and maintenance costs were paid by the Exchequer.[112]

Although prices had risen through inflation, increased royal revenues allowed these ships to be built. Aylmer cites rising annual five-year averages for selected revenue sources, producing £618,376 in 1631-35 and £899,482 in 1636-41. He attributes this to increased tax farm rents and rates on taxed goods, leased timber rights in the Forest of Dean, more efficient administration and some improved trade. Fines for not taking up a knighthood were an important extra-parliamentary source.[114]

Sovereign of the Seas and the rebuilt *Prince Royal* were clearly much larger than the other ships built in the 1620s and '30s. In building *Sovereign of the Seas* Perrin considered that Charles initiated the idea of a 'royal ship that should be larger and more ornate than any of her predecessors' in June 1634. 'Being informed that his Majesty is minded to build a great ship of these dimensions', Trinity House opposed a ship of this length (124 ft) and draught (22 ft) because no English port, apart from the Isle of Wight, could 'in safety harbour' her, and two ships 'of more force' could have been built for this money. However, Charles approved Chatham master shipwright Phineas Pett's design in April 1635 for a 'new great ship' of 1,522 tons. Pett estimated its cost at £13,860, but it totalled £40,833. The keel was laid in December 1635 and it was launched and named on 14 October 1637. The rebuilding of *Royal Prince* cost £17,450.[115]

Revenue from crown lands fell progressively during the 1630s because many estates had been sold in the late 1620s. By 1640 most future revenues had already been mortgaged for one, five or even eleven years. By 1640 Charles had little ready money and no credit. This hampered the Navy Board's ability to manage the navy effectively as *matériel* and services were more expensive when payment could not be made within a customary period or without ready cash, as Pepys also noted.[116]

Hollond, an officeholder in 1638, argued that the business of ports had increased and that English ships were protecting Dutch and French merchant ships.[117] Oppenheim, however, considered that although English mercantile activity increased in the 1620s, the Dutch were overtaking the English in merchant and naval shipping and fisheries; France was wealthier than England; and north African corsairs and Dunkirk privateers raided Irish and English coasts from Newcastle to Bristol with impunity until Rainborough's expedition against Sallee in 1637 quelled the former attacks. Only Spain was no longer a threat.[118]

Pennington was critical of the sailing qualities of most of the new built ships; ten prizes gained in the 1620s and 30s were useful additions to the fleet. He reported in 1634 that the Dutch and pirate hulls were cleaned and tallowed every two or three months, 'which makes them go and work better than ours', whereas English ships were cleaned two or three months before sailing, but not tallowed, and kept out for eight to ten months, so overgrown with barnacles and weeds. English ships had one advantage in that they were built more solidly, therefore endured being shot at more than the Dutch.[119] He recommended that *Swan*, a Dunkirker captured in 1636, should be used as a model by English builders.[120]

Finally, chips, the right of dockyard workers to wood waste allowed abuse. Chips were small, thin pieces of wood separated by hewing, made by the woodcutter in the course of their work, recorded from 1330. The necessity of clearing the dockyards of waste wood would have been a waste management policy and fire precaution in any permanent shipyard.[121] In 1625 at Chatham Dockyard the perquisite was extended from carpenters in the ordinary 'to caulkers, housecarpenters, joiners '& their wifes and children', who collected them and split larger pieces of timber every Saturday.[122] Joshua Downing, Assistant Navy Commissioner at Chatham, 1625-1628, recommended that Carpenters in the Ordinary be financially compensated, to save loss of time and destruction of timber, and they were allowed 1d a day.[123] However, these payments were not maintained after 1626, and by 1634 men were carrying out wood three times a day.[124] Hollond describes the waste of timber: 'cleaved for firing, embezzled for building or repairing private houses'. On 15 November 1634 the Admiralty ordered the perquisite 'to be distributed according to the ancient order established, and that in lieu of the same there be allowed to the master shipwrights and chief of the workmen the old allowance of one penny a day'.[125] Oppenheim cites a lighter containing 8,000 treenails stolen from Deptford for a government shipwright who also owned a private shipyard.[126] In 1702 Commissioner Henry Greenhill filled a lighter with chips as an experiment at Deptford.[127] In 1673 and 1698 Greenhill reported 'sev:ll abuses practiced in y:e carrying of Chips out of his Maj:ties Yard att Woolwich and Deptford to his Maj:ties great detriment', the workmen preparing 'both Lawfull and unLawfull Chips to be exported by theire wives Children or Friends upon the accustomary dayes'.[128]

In March 1627 Coke sent the Danish Andersen to visit Chatham. He found that money was lacking for stores to make repairs, the men had not been paid for 18 months and materials were being stolen. In January 1630 a storm damaged storehouses, which could only be repaired by selling old cordage. However, £2,197 was spent in 1629 on Portsmouth, Deptford and Chatham, and a further £2,445 in 1634. Woolwich declined, part being leased to the East India Company in 1633 at £100 a year, used to build a wall around the yard and repair buildings. Before Buckingham died he had ordered estimates for a new double dock at Portsmouth to replace the first one destroyed by storms. In 1630 principal officers Phineas Pett, Sir Thomas Aylesbury and William Burrell visited Portsmouth to inspect its capabilities. They reported that there was 'no worm destructive to ships bred in Portsmouth harbour'. More ships were moored there because of its convenience for campaigns, and more shipwrights were sent in 1645. Edward Boate was master shipwright from 1638.[129] In 1634 'Chatham was now the first of English dockyards' and contained 70 or 80 acres, with a brick wall around part of it. By 1637 the storehouses were better supplied, thanks to ship money, although the principal officers desired more timber and cordage.[130]

Bottom-up Correspondence

The following Navy Board correspondence to the Admiralty presents a bottom-up perspective. It is necessarily one-sided, although Admiralty instructions or attitudes are sometimes indicated.

In January 1631 William Russell, Kenelme Digby and Dennis Flemyng reported

that His Maj:ties Ships att Chatham continue still with their Rigging up, which makes them more Subject to the Stress of Wind and weather and their Ground Tacle being much worne by long lying

They requested an Admiralty direction that the rigging be taken down, 'whereby the Ships may ride with more safety' or a 'Warrant for the continuance thereof'.[131] They also suggested additional ships' carpenters '(not provided for in the Estimate)':

to be constantly residing either on board the great Ships, or so near att hand as they may be employ'd upon all suddaine actions, that may fall out, to stop a Leake, opening and Shutting of the Dock Gates, and att the bringing in or Launching of the Ships, as often as they be graved or trimmed in dry Dock, & foreseeing an apparent necessity of the continuance of about 34 Shipwrights constantly in the workes, who may be well employ'd in making of masts repairing of Boates, and other small Defects that dayly happen on board the Ships, besides their Labour att the Season of the Year in Caulking time, who may deservedly earn their Wages, being Muster'd by Prick and Cheque as other men employ'd on day wages are wont to be[132]

This would not involve an increase of the annual ordinary naval estimate because they proposed removing two shipkeepers from each of the first and second rank of ships and one in the third rank and entering twenty Master Carpenters and fourteen apprentices in their place who would sleep on board the ships. They proposed that the apprentices be paid as shipkeepers, and the master carpenters to have the

> *overplus of their Wages borne, out of the monys allotted for Shipwrights Pforming the ordinary repairations, which is no new thing though in another kind, for that for many years itt hath been usuall to take a certain Select Company of Shipkeepers to worke day labour in the Dock and held good Thrift, But with your Lo:^{ps} favour wee hold these men of much more use* [133]

These extracts show firstly that the Navy Board aimed to manage the resources of the navy more effectively, and secondly that it was initiating policy at the bottom level of administration; requesting direction to stop rigging being misused and recommending that skilled carpenters should be permanently on the Ordinary payroll to effect timely repairs.

The cost and inconvenience of emergency repairs was highlighted when the Navy Board reported that 'the Lyons Whelpe, now Riding att Hunger Road near Bristoll', had 'sundry defects of Carpenty [sic] Workes in the said ships Hull', and needed to be 'carreen'd and Supplyd with a Boate and Divers other necessaries SeaStores, before She goes forth againe to continue her employm.^{t}'. This would take two months and also cost food for the crew.[134] By 30 May 1632 repairs necessary to the 5th and 9th Lyons Whelpes at Bristol amounted to £968 1s 8d but only £426 had been sent from the Lords Justices of Ireland. Mr Kitchen, Master and Brother of Trinity House, who had overseen their careening and fitting, was owed £542 1s 8d. They warned the Admiralty that 'it would be a great needless charge to keep 120 men Idle for want of some necessaries to sett them forth'. They ended by reminding the Admiralty that Sr Wm Russell was still owed £331 1s 11d 'for fitting the said Ships to Sea the last Year'.[135]

Timber provision and carriage was a dominant issue. In April 1632 the Navy Board informed the Admiralty that Peter Pett,[136] 'one of His Maj:^{ties} Ma^{r} Shipwrights employed in making Choice and Markeing of Timber, in the Forrests of East and West Bear and the new Forrests in the County of Southtōn', had selected 2,000 'good and Serviceable Trees to be Marked in ye new Forrest (part whereof about the Numb^{r} of 600 are already felled)'. The 2,000 trees were 'Estimated to amount to 3000 Loads'. However, in the

> *Forrest of East and West Beare he findeth only 400 Trees fitt for Service, and those only for Treenailes and Short Planke wch will not make above 400 Loads, So that there is wanting 1600 Trees of the N:^{o} expected from that forrest, which may be Supply'd in the new Forrest, or in the forrest of Als=Holt^{137} in the County of Surry (as he informeth us) where he findeth the Timber generally good* [138]

They explained that 'this Provision of Timber (for the most part) is made for the Ships intended to be built the next year', and suggested, if the plans could be resolved by the master shipwrights, that the

> *Timbers should be Moulded in the Woods,^{139} which would ease both King and Country near one third part in the Land & Waterige thereof, the Quantity of Timber wanting well knowne and the Cumbrance of Superfluous unnecessary Stuff in His Maj:^{ties} Timber Yards avoided ~ And considering that the burthen of Land Carriage of so many Loads will lye very heavy upon that County, if it should wholly be charged upon them, Wee thought it not unfitt to represent unto your Lo.^{ps} on their behalf (to the end His Maj:^{ties} Service may be Pformed withal chearfullness) that you would be pleased to take some Order, for easeing their Charge, by joyning such Numbers of the Adjacent Counties for performance of this Service as shall seeme fitt as in the like case hath been formerly don Joyning ye Counties of Berke and Buckingham with the County of Oxon in Carriage of the Provision of Timber, made in his Highness's Forrests of Shottover and Stowood* [140]

Again, the Navy Board, from its expertise, was suggesting policy to save money. It was also acting as a mediator between county inhabitants and the king to save carriage charges, suggesting that several counties should bear this cost. The reason was pragmatic: it aimed for 'chearfullness' in transporting timber for the navy, otherwise the process would take much longer or might not occur at all. It proposed that the carriage of 3,400 loads of timber could be divided between Wiltshire and Sussex, and urged that 'The Season of ye Year (now fitt for carryage) will admitt of no Delay'.[141] On 25 July it urged Admiralty Secretary Nicholas to send a warrant for

> *felling 1600 Trees in Alsen-Holt and promised by Mr Lake to be perfected to morrow in regard the Yeare is so farr spent wee desire to husband the Remaind^er to the best advantage wee can and to P^rpare all things in readiness to putt it in speedy Execution*[142]

The Navy Board, driven by seasonal exigencies, is driving the Admiralty Board to direct the harvest of timber urgently required for the following year, confirming that unseasoned timber was being used.

In 1633 the Navy Board reported resistance to the traditional services of land and water carriage of 'Ship Timber and Planck appointed for the new building of His Maj:^ties Ships now in hand in His Highness Dock Yards att Deptford &Woolwich' from Shotover and Stowood.[143] Although this service was for 'the defence and Safety of the Kingdome', inhabitants in Cheam, Ewell, Windsor, Guildford and Oxford claimed exemptions and 'refused to send their Carts' to carry timber. On the River Thames barge owners refused to offer their barges because they would lose money as the king paid so little for purveyance, and Berks, Surrey and Dorset Justices of the Peace were not enforcing royal directions for carriage.[144]

The Board also urged the Admiralty to authorise the manufacture of cordage, as the navy had 'increased more than a third part in the last twelve years'. Production in the yards would prevent reliance on merchants and ensure good workmanship and strength if made 'in the dayly view of the men who hazard their lives in the Ships att Sea'.[145]

In January 1633 the Navy Board alluded to the 'Extraordinary waste carried out by the Shipwrights, Sawyers, House Carpenters, Labourers and other workmen three times a day: at breakfast, dinner and night under colour of chips'. They note: 'Much good timber is cut into short ends and cleared into chips, Portable under their arms.' They hide 'Imbezellings' amongst these chips in their cabins. A lighter was discovered with 3 or 4 loads of Treenailes which the shipwrights claimed had made from chips, to be carried out of the Yard. The cutting of chips was also a waste of time. But 'these men are a clamorous people, hardly to be broken from their ancient liberties.' The Navy Board recommended removal of the perquisite: the larger chips to be collected for use in the Pitch Kettle; the smaller chips to be given to the poor widows and children of the parish, a measure which could not be rejected by the shipbuilding community. To compensate for the loss, it was proposed that the shipwrights should be paid twopence a day, 'to prevent that overgrowne mischief herafter'. They desired a swift response from the Admiralty because 'the two new ships are now in hand, and they are not to be restrained from their old way, without an order from his Majesty'.[146]

Conclusions

Evidence from these secondary and primary sources indicates that many of the restoration naval administrative structures, principles, methods and relationships derived from the sixteenth and early seventeenth centuries. The restoration therefore represented continuity from the reigns of James and Charles, plus changes from the interregnum which were not always new. James as Lord High Admiral promoted 'collective administrative and financial responsibility' and 'the deliberate use of one officer to duplicate the work of another, in order to keep a check on the former's honesty and efficiency'.[147]

As within all government departments, family networks provided routes for advancement. The dockyards and officers comprised an extended family connected by marriage or blood, but also riven by jealousy and greed.[148] Seeking to explain later political allegiances, Aylmer concludes that by 1642, benefiting from court favour, 'some of the King's servants were beginning to be the public servants of the Crown while others were becoming more specifically the private servants of the sovereign.' This analysis does not address the issue of duty, as opposed to loyalty. He asserts that 'a fundamental reform of the revenue and spending departments, and of the methods of administration...was far beyond the capacity of Charles I's régime'. He distinguishes between 'full-time professional civil servants' and those who were also 'landed gentlemen, practising lawyers, or...actively engaged in commerce.' The period itself did not perceive 'civil servants' as an occupational classification, but they formed a 'coherent social class with common traditions, interests, and aspirations', belonging or aspiring to the landed armigerous classes. They were 'his' servants, but he had not appointed all of them. Some were 'personal', some were royal, servants.[149]

Charles I's navy and many of its personnel were driven by Elizabethan aspirations of national defence and offence which could be achieved occasionally by privateers but were now operating within a different Europe. The ten *Lion's Whelp* pinnaces of the Ship Money fleet were nimble but too small and lightly gunned to protect trade, while the larger aging Elizabethan ships were cumbersome. *Sovereign of the Seas*, with three decks and 102 guns was the most powerful fighting ship in the world, effective in the First Dutch War, but did not see action under Charles. By 1639 the fleet was failing to enforce sovereignty of the seas, even in the Channel. Ship money could not fund the dockyard and ordnance infrastructure, which was financed by the Exchequer whose general income was £620,000 in 1635.[150]

Without national consensus on the navy's purpose, funding and leadership, comprehensive merchant and coastal protection and a more unified bureaucracy with enforced duties, naval administration could not achieve the aims of both Charles and the nation as expressed through parliament. John Hampden's lawsuit against the legality of Ship Money was defeated in 1638, but the Act against levying Ship Money received royal assent on 7 August 1641.[151]

The 1630s were a cuspic period, mixing traditional structures, aspirations, images and icons. Charles's struggle to realise his aims alone led to civil war, which became a catalyst for investing even greater material wealth in the navy.[152] Ironically, £26,500 of the proceeds of the sale of Charles I's art collection (£134, 383 5s 4d) addressed 'the present necessaries of the navy' in 1649. The statue of Charles, which had been sold to a brazier in 1650 to be broken up, 'that nothing might remain in memory of his said majesty', reappeared in 1660, bought by Charles II and erected in 1676 at Charing Cross, facing Whitehall and the Banqueting Hall where he had been executed in 1649.[153] Charles I affronted and alienated local government, his necessary supporters, a feat repeated by his son James II, who did not learn from history.

Puritan principles, naval abuses and financial exigencies shaped both the 1630s and the 1660s but the essential dichotomy of salaries versus fees, reversed somewhat during the interregnum, was again transposed: the restoration comprehensively restored fees as well as Charles.

References

1. R C Latham and W Matthews, (Eds). *The Diary of Samuel* Pepys, volume II, 1662, Harper Collins, London, 1995, 145; *The Diary of Samuel* Pepys, volume IX, 1668-9, Harper Collins, London, 1995, 489; J R Tanner, Ed., *Hollond's Discourses of the Navy 1638 and 1659*, NRS, London, 1896, ix-xviii, liii, liv, 99. George, born 1623, was Joshua's nephew, son of Emmanuel b. 1585, brother to Joshua.

2. R Latham, Ed., *Samuel Pepys and the Second Dutch War*, Navy Records Society, Aldershot, 1995, 385.

3. Tanner, 1896, xiv-xvi

4. G E Aylmer, *The King's Servants: the Civil Service of Charles I, 1625-1642*, Routledge & Kegan Paul, London, 1961, 77, 124, 261, 337-344, 393-421, 467, 481.

5. The National Archives (TNA), ADM 106/3, fo 212; ADM 106/2, 21.9.1660; ADM 2/1745, fo 9v.

6. M Oppenheim, *A History of the Administration of the Royal Navy and of Merchant Shipping in Relation to the Navy 1509-1660*, Bodley Head, London, 1896, 286.

7. British Library (BL), Add. MS 9301, Navy Reports 1618-1637.

8. Tanner, 1896, liii; A P McGowan, *The Jacobean Commissions of Enquiry 1608 and 1618*, NRS, London, 1971, xiii.

9. McGowan, xiv, 11-19; N A M Rodger, *A Naval History of Britain, I, 660-1649: The Safeguard of the Sea*, Harper Collins, London, 1997, 364-8, 508-9.

10. McGowan, xiii; Aylmer, chapter 3, especially 69, 75-6, 177-8.

11. Aylmer, 179, 189, 192.

12. Linda Levy Peck, *Court Patronage and Corruption in Early Stuart England*, Routledge, London, 1993, 123; Latham and Matthews, I, *1660*, cxix; IV, *1663*, 201, 409, 415; V, *1664*, 141.

13. Oppenheim, 1896, 299; TNA, ADM 106/3539 part 2, *c.*1652-1653.

14 Aylmer, 175-8, 192, 249.

15 Aylmer, 76.

16 J P W Ehrman, 'The official papers transferred by Pepys to the Admiralty by 12 July 1689', *The Mariner's Mirror*, 34:4, 1948, 255-270; R D Merriman, Ed., *The Sergison Papers*, Navy Records Society, London, 1949, 2; M Lower, 'Some Notices of Charles Sergison Esq', *Sussex Archaeological Collections*, XXV 1873, 62-84; A Thrush, 'The Government and its Records, 1603-1640', History of Parliament Trust, retrieved 3.1.2012 from http://gale.cengage.co.uk/images Thrush%20Government%20Records%201603_1640.pdf; A Coats, 'The Economy of the Navy and Portsmouth. A Discourse between the civilian naval administration of Portsmouth Dockyard and the surrounding communities, 1650-1800', DPhil Thesis, 2000, University of Sussex, 43-4. Pepys's personal and official papers are in the Rawlinson Collection, Bodleian Library, Oxford. His library of 3,000 volumes and 250 volumes of manuscripts were bequeathed to Magdalene College Cambridge at the death of John Jackson, Pepys' nephew and heir, in 1724.

17 Aylmer, 106, 217-20, 227-233; 275-7.

18 *Ibid.*, 209, 211, 245.

19 Adapted from Aylmer, 205, 207-9, based on Exchequer Fee and Pension Lists.

20 Aylmer, 42, 106, 211.

21 *Ibid.*, 136, 192, 209, 229-30, 352.

22 *Ibid.*, 338-9.

23 *Ibid.*, 227-3.

24 Young, 69-70.

25 Aylmer, 137, 176, 181, 186, 193-5, 204-210, 233-4.

26 *Ibid.*, 62-3.

27 *Ibid.*, 122, 364-5.

28 Reversions, a written grant of succession to an office or group of posts, made in the king's name, were made normally by whoever held the gift of appointment, but an influential person could interpose. 'During good behaviour' was more secure than 'during the king's pleasure' because misbehaviour in office had to be proved in court. Aylmer, 108, 110, 112, 122-5.

29 *Ibid.*, 260-3, 272-3.

30 Rodger, 391.

31 McGowan, xix, xxvi; Aylmer, 78, 81, 408, 409.

32 Aylmer 195.

33 Remembrancer: a keeper of rolls and registers who audited revenue accounts in the traditional way, evolving from the 12-14th centuries. *Compact Edition of the Oxford English Dictionary*, 2 volumes, Oxford University Press, 1986 (*OED*); Aylmer, 34.

34 Gofton died in 1628. His servant was George Bingley, with no family influence, later Auditor of the Prests *c.*1635-42. Aylmer, 78.

35 Prest: a loan (sometimes forced), advance of money to the crown or purveyance whereby the crown obtained goods or services at a fixed rate. *OED*.

36 McGowan, xxvii.

37 Aylmer, 91.

38 M B Young, *Servility and Service. The Life and Work of Sir John Coke*, Boydell/RHS, Woodbridge, 1986, 43.

39 Young, 43-7, 56. Young quotes the Lord Keeper, Bishop Williams, as observing that Cranfield was 'not pryvye to those imployments havinge been otherwise bredd, used the helpe of others, namely of Mr. Cooke, and at last assumed the wholl glory to himselfe'. This is an interesting aside on their relative status.

40 Young, 64-9, 70.

41 Aylmer, 360, n. 3: 'The younger Vane; William Batten, "furious in the new fancies of religion", Clarendon, *Hist, Reb.*, I, 225; Thomas Smith, "highly infected with Presbyterian principles" (P. Warwick, *Memoires of the Reigne of King Charles I*, 1701, p.120); Thomas Barlow (see refs in Pepys's Diary for 1660-5).'

42 Aylmer, 77-8, 135, 374-6.

43 *Ibid.*, 132-4, 419.

44 *Ibid.*, 175.

45 Young, 47, 50-1; McGowan, 265.

46 McGowan, 297.

47 *Ibid.*, 273-4.

48 *Ibid.*, xvii-xxvii.

49 *Ibid.*, xvii.

50 Young, 48.

51 McGowan, xvii; BL, Sloane MS 3232, fos 139, 139v.

52 Quoted, Peck, 118-9.

53 Young, 54-6; Legatt, Chatham clerk of the cheque, was granted the post of clerk of the navy in reversion after Peter Buck 17 April 1604, but did not succeed. A provisional list compiled by J C Sainty, January 2003, Institute of Historical Research, retrieved 16 April 2010, www.history.ac.uk/resources/office/navyclerk.

54 McGowan, 284-5.

55 Tanner, 1896, 101, xliii. Clenching: grasping.

56 Oppenheim, 1896, 193-4.

57 Oppenheim, 1896, 195. Dead ships: non-existent ships' names were carried on the books.

58 McGowan, 286, 287, 289, 296-8.

59 Young, 50, 55.

60 Dead pays: pay continued in the name of dead or discharged seamen, appropriated by officers or officials. Peck, 122, 123, 125.

61 Young, 53-4.

62 Young, 55.

63 Rodger, 369; Young, 49, 75; Oppenheim, 195.

64 McGowan, 287, 288, 296.

65 Young, 56-7; McGowan, xvii-xxvi; Rodger, 369, 507.

66 Rodger, 481-2; Young, 75.

67 Rodger, 481-2; Young, 75, 78; J G Coad, *The Royal Dockyards 1690-1850*, RCHM/Scolar Press Aldershot, 1989, 90, 117, citing E Hasted, *The History and Topographical Survey of the County of Kent*, 1797, 2, 344.

68 Young, 75-87. Ordinary: operational and reserve navy and personnel funded by annual naval estimates.

69 Tanner, 1896, 41, notes 2, 3.

70 Young, 75-83.

71 Rodger, 357-9; Young, 135-7, 186-9.

72 Rodger, 359.

73 N Matar, *Britain and Barbary 1589-1689*, University Press of Florida, Gainesville, 2005; N Matar, 'The Barbary Corsairs, King Charles I and the Civil War', *Seventeenth Century*, 16:2, 2001, 239-58, accessed 28 March 2010, www.manchesteruniversitypress.co.uk/uploads/docs/160239.pdf. Note that corsairs and pirates are not always synonymous.

74 Oppenheim, 1896, 196-7, 205.

75 *Ibid.*, 210

76 McGowan, xvii, xix, xxvi; Rodger, 481.

77 Rodger, 370-1.

78 Oppenheim, 1896, 219-25. 13 belonged to the king and 20 were Dutch. Rodger, 357-9.

79 Aylmer, 136, 179, 192-3, 229-30.

80 Rodger, 371; Young, 141-8, 164, 167, 171-185.

81 Oppenheim, 1896, 292-3.

82 Ibid, 225-30, 259-60.

83 Rodger, 360, 374; Young, 189-90; D. Dymond, *Captain John Mason and the Duke of Buckingham*. Portsmouth Paper 17, Portsmouth City Council, 1972, 9-10.

84 Dymond, 12-16; Oppenheim, 1896, 227-236.

85 Rodger, 360-1; Young, 189-204.

86 Aylmer, 42; Oppenheim, 1896, 279-80, 282. Wallingford House had been purchased by the Duke of Buckingham in 1622 and remained in his family. It was demolished in 1694 and the first

Admiralty Office was built on the site 1694-5. 'The Admiralty', *Survey of London: volume 16: St Martin-in-the-Fields I: Charing Cross*, 1935, 45-70: http://www.british-history.ac.uk/report.aspx?compid=68108, accessed 2 January 2012.

87 Rodger, 391, 507-9; Young, 134; Aylmer, 61, 89, 92, 110-11, 115-6, 132-3, 195, 360, 365, 468, 480; TNA, E403/2563, p.1; Oppenheim, 279; 'September 1642: Ordinance appointing Tho. Smith Secretary of the Admiralty', *Acts and Ordinances of the Interregnum, 1642-1660* (1911), 29-30: http://www.british-history.ac.uk/report.aspx?compid=55743, accessed 2 January 2012; K R Andrews, *Ships, Money, and Politics: Seafaring and Naval Enterprise in the Reign of Charles I*, Cambridge University Press, Cambridge, 1991, 188. Nicholas served Edward de la Zouch, Warden of Cinque Ports during James's reign, transferred to George Villiers when he succeeded him and after serving other courtiers, to Charles, Prince of Wales. Aylmer 78-9, 130-1, 132-3, 142-3.

88 J Brotton, *The Sale of the Late King's Goods. Charles I and his Art Collection*, Macmillan, London, 2006, 42-50, 111, 158, 160-1, 348-51. Also see K Sharpe, *Image Wars: Promoting Kings and Commonwealths in England, 1603-1660*, Yale, 2010.

89 Rodger, 79, 99, 114,150, 380-3, 388. The government issued John Selden's *Mare Clausum* in 1635 to defend the doctrine.

90 Oppenheim, 1896, 252.

91 Rodger, 362, 374, 391, 392, 508-9; McGowan, xvii; Oppenheim, 1896, 195, 280, 281-2; Tanner, 1896, 19, n. 1; Aylmer, 77-8, 81, 86, 91, 93-4, 102, 163, 166-8, 231, 321, 360, 365, 382, 481; McGowan, 184, 163, 187;Young 74; W L Clowes, *The Royal Navy. A History From the Earliest Times to 1900*, II, Chatham Publishing, London, 1996, 16-7, 19, 71; BL, Add. MS 9301, Navy Reports 1618-1637; A provisional list compiled by J C Sainty, January 2003, Institute of Historical Research.

92 Pepys was acting in the post before he received his patent, sealed by the Lord Chancellor on 13 July 1660. The patent, costing him £40, revoked the life appointment of Thomas Barlow, 'an old consumptive man and fair-conditioned'. On 23 July he signed an agreement to buy out Barlow for £50 a year until his death (in 1665) or £100 if there was a salary increase to £350, which occurred on 7 July, to avoid a challenge to his position. Latham and Matthews, I, 188-206; J Collinge, *Navy Board Officials, 1660-1832*, University of London, 1978, 22.

93 J R Tanner, Ed., *A Descriptive Catalogue of the Naval Manuscripts in the Pepysian Library*, I, NRS, 1903, I, 19-20. For a list of posts in Charles I's reign, see Aylmer, 480-1.

94 McGowan, xiii, 273, 296, 297.

95 Oppenheim, 1896, 190.

96 Rodger, 371; BL, Add MS 9301, fo 17.

97 Clowes, 19. See T F Reddaway's maps of London in the 1660s in R C Latham and W Matthews, Eds, *The Diary of Samuel Pepys*, volume X, *Companion*, Harper Collins, London, 1995, 632-5.

98 Aylmer 76, citing M Oppenheim, Ed., *Naval Tracts of Sir William Monson*, III, NRS, London, 1912, 416-8. Monson was Admiral of the Narrow Seas (English Channel) 1604-16. Rodger, 352.

99 Aylmer, 78, 93-4,132-3, 135, 166-8, 419.

100 McGowan, 260-1.

101 Oppenheim, 1896, 280-84, 85-6.

102 Rodger, 391.

103 Tanner, 1896, ix, xii; Aylmer, 350.

104 Rodger, 392-3.

105 Rodger, 375-8, 391-2, citing J J Keevil, *Medicine and the Navy 1200-1900*, volume I, 1200-1649, Livingstone, Edinburgh/London, 1957, 179.

106 Oppenheim, 1896, 285-6.

107 Tanner, 1896, 74.

108 *Ibid.*, 79-80; BL, Sloane MS 3232, fos 139, 139v.

109 *Ibid.*, 81-5.

110 Tanner, 1896, 81-100.

111 Aylmer, 7, 65, 67, 113, 154; Rodger, 381-2,

112 Rodger, 382-3, 385, 391.

113 Rodger, 388, 482; Clowes, II, 11; Oppenheim, 1896, 254-5. Oppenheim states that *Roebuck* and *Greyhound* were built from the waste of *Sovereign*.

114 Aylmer, 63-7, 175-6: Great and Petty Customs Farms, New Impositions (Customs), Additional New Impositions, Coal Duties and Farms, Payments from Soap-makers, Forest of Dean, The Mint, Court of Wards and Liveries. Rodger, 315-6.

115 Perrin, 1917, xci, xcii, xcviii, 166-7, 214- 6; Rodger, 388. Rodger, 388, states that the cost of *Sovereign of the Seas* was £65,586 16s 9½d.

116 See Tanner, 1896, 71, n. 1.

117 *Ibid.*, 1896, 5-6.

118 Oppenheim, 1896, 198-201, 217-8, 272, 274-8. The French were occupied by the Fronde and generally avoided the ship money fleets.

119 Oppenheim, 1896, 253-4.

120 *Ibid.*, 254-8.

121 *OED*. The Princess Yachts planning application in South Yard, Plymouth (2010) included a policy for reusing wood waste. Design & Access Statement for Princess Yachts, Devonport, Plymouth Rev A2 05/05/2010, accessed 5.1.2012, http://www.plymouth.gov.uk/planningdoc-2?appno=10-00640-FUL

122 A D Thrush, 'The Navy under Charles I 1625-1640', PhD thesis, University of London, 1991, 119.

123 Tanner, 1896, ix, 99.

124 Thrush, 119-121.

125 Tanner, 1896, 96-7, n. 1., citing *CSPD*, 1634-5, 293.

126 Oppenheim, 1896, 285.

127 A Coats 'Breakfast and chips - signifiers of power relations in dockyards. Naval shipbuilding at Deptford and Woolwich dockyards', *Papers of the Second Symposium on Shipbuilding on the Thames and Thames-built Ships*, R Owen Ed., 2004, 111; TNA, ADM 106/555, 22 September 1702.

128 National Maritime Museum, POR/A/101, fo 168, 17.10.1698, Commissioner Henry Greenhill to the Respective Officers at Portsmouth Yard.

129 Oppenheim, 1896, 296-7; Coats, 2000, 12, 352; BL, Add. MS 9301, fo 33v. *Teredo navalis*, shipworm, was brought from warmer waters and had infected Sheerness dockyard.

130 Oppenheim, 1896, 297-8.

131 BL, Add. MS 9301, fos 23v, 31 January 1631, to Edward Nicholas, Admiralty Secretary. This contrasts with the Master Attendants' recommendation to the 1618 commission that 'Every ship should be fully and completely rigged so that the ships might be ready for any service.' McGowan, xxii.

132 BL, Add. MS 9301, fo 23v-24v, 31 January 1631, to the Admiralty from Russell, Slingisbie, Pett and Flemyng.

133 *Ibid.*.

134 *Ibid.*, fo 27v, 19 February 1632, to the Admiralty from Russell, Digby, Edisbury and Flemyng.

135 *Ibid*, fo 27v, 19 February 1632, to the Admiralty from Russell, Digby, Edisbury and Flemyng.

136 Peter Pett (1610-1674), brother of Phineas Pett, a private shipbuilder and a purveyor, referred to with Richard Merritt Sr in 1608 depositions. McGowan, xv, 55, 70, 176, 232, 243-4.

137 See remarks about Alesholt in 1608, McGowan, 52-6, 242, 253, 254. Alice Holt, south of Farnham in Hampshire, noted for its oaks, used extensively for shipbuilding in the Tudor and Stuart period and depleted by 1635.

138 BL, Add. MS 9301, fos 29-29v, 30 May 1632, to the Admiralty from Russell and Digby.

139 See the 1618 Commission's reference: 'Crooked timber to be moulded in the woods'. McGowan, 304, 242.

140 BL, Add. MS 9301, fos 28v-29, 13 April 1632, to the Admiralty from Russell, Palmer, Digby and Flemyng. See McGowan, 221, 253 for Stowood and Shotover being used by the navy in 1608. Timothy Tyrrell the elder was appointed Steward of Shotover and Stowood in 1613, when master of the buck hounds to Charles, Prince of Wales. Charles I extended the bailiwick for the lifetime of Sir Timothy's son, Timothy the younger. 'In 1632 there is the significant statement that owing to decay of timber in the Forest of Dean, the New Forest, and Waltham Forest, Shotover and Stowood were the only sources of supply. Due to depredations by the Tyrrells: 'In 1631, of 14,000 oaks marked for the navy (4,000 for present and 10,000 for future use), many were alleged to be of little worth. Further, there were difficulties about obligation to transport the timber from the forest. It was held that neighbouring counties should help with this work, as they did in the defence of the realm, and appeals were lodged against the duty. In 1635 merchants were offering tenders to do the work.' In 1660, before disafforestation by Charles II, it was 932 acres in extent. 'Parishes: Shotover', *A History of the County of Oxford: Volume 5: Bullingdon Hundred*,1957, pp. 275-281: http://www.british-history.ac.uk/report.aspx?compid=101898 Date accessed: 31 December 2011. Charles I attempted to revive forest

jurisdiction in the 1630s. 'The officers of the Navy certified to the lords of the Admiralty, in August, 1636, that 200,000 tree-nails would be required for His Majesty's yards, and for the repairing of the *Anne Royal*. It would require 1,000 young trees for that number of nails to be cleft out of the heart; the best trees for the purpose were to be found in the forests of Shotover and Stowood. The king, however, in the following month, declined to allow such an extensive felling at Shotover, and eventually the tree-nails were obtained from the New Forest.' *CSPD* 1636–7, 104, cccxxx, 71; 137, cccxxxii, 18. 'Forestry', *A History of the County of Oxford, Volume 2*, 1907, 293-301: http://www.british-history.ac.uk/report.aspx?compid=101947, accessed 31 December 2011.

141 BL, Add. MS 9301, fo 29, 13 April 1632, to the Admiralty from Russell, Palmer, Digby and Flemyng.

142 *Ibid.*, fo 29v, 25 July 1632, to the Admiralty from Russell, Palmer, Digby, Flemyng and Pett.

143 Ibid, fos 31, 34-5.

144 *Ibid.*, fo 32.

145 *Ibid.*, fo 31.

146 *Ibid.*, fo 32, Palmer, Edisbury, Flemyng.

147 TNA, ADM 7/633, Duke of York's Instructions.

148 Aylmer, chapter 3, especially p.81.

149 *Ibid.*, 461-7, 420, 421.

150 Rodger, 386-91, 412.

151 Rodger, 393; 'An Act for the declaring unlawfull and void the late proceedings touching Ship money and for the vacating of all Records and Processe concerning the same', August 7, 1641, 17 Car. I. cap. 14, *Statutes of the Realm: volume 5: 1628-80* (1819), 116-117, accessed 30 December 2011: http://www.british-history.ac.uk/report.aspx?compid=47225.

152 Rodger, 433.

153 Brotton, 210, 307, 310, 348-51.

Dr Ann Coats *is secretary of the Naval Dockyards Society. She is at the University of Portsmouth.*

PARLIAMENT, ROYAL DOCKYARDS AND THE LONDON MARITIME COMMUNITY: THE AFTERMATH OF THE 1648 NAVAL REVOLT*

Richard J Blakemore

Abstract

The background to the second civil war of 1648 is outlined and the issue of the royal dockyards together with their place in the maritime community examined. The events of 1648 are considered in detail, centring on the contest for control of the dockyards, on the loyalty of dockyard workers and on the role of the maritime community itself. It is noted that the 1649 purge of those who had supported the royalist cause might have secured the dockyards for parliament, but did little to ensure a lasting solution.

Introduction

In May 1648, during the various regional conflicts known collectively as the second civil war, a large proportion of the Royal Navy (which had in fact been under parliamentary control since 1642) rejected their commander and eventually joined the Prince of Wales to fight against parliament's forces. At the same time, royalist outbreaks in Kent and Essex turned control of the naval dockyards on the Thames into a vital, contested issue. Attempts were made during May by the Kent royalists to take control of the dockyards and stores at Chatham and Deptford, although these were defeated by parliamentary forces. Even so, in August the Earl of Warwick, parliament's admiral, issued warrants to arrest any dockyard workers who had been involved in 'the late Rebellion in Kent', indicating considerable doubt over the loyalty of the maritime community.[1]

This paper aims to use the events of the summer of 1648 to understand the importance of the Thames dockyards both to the central government and Navy, and within the maritime community. Royal dockyards played an important role in the relationships between these various groups, both as social focal points, and as centres of state activity, which rendered them sites of potential political volatility. The contest for control of these focal points, and parliament's eventual victory in that contest, was an influential factor in a crucial phase of the English revolution, allowing the new model army's allies to raise and supply a fleet, thus contributing to their victory in the second civil war. Control of the royal dockyards probably also served a significant symbolic purpose in the struggle for the loyalty of the maritime community which ensued in the summer and autumn of 1648.

The Origins of the Second Civil War, 1646-8

The conflicts of 1648 were the result of widespread dissatisfaction with the parliamentary regime following the surrender of Charles I in 1646, which ended the first civil war. The lack of progress in negotiations with the king, the continued existence and influence of the new model army (and the taxes levied for its upkeep), and the lack of religious settlement, all contributed to parliament's growing unpopularity.[2] Whereas parliament and its supporters had been united during the first civil war by the need to defeat the king, in the wake of victory they struggled to find a consensus on a variety of issues, and the parliamentary leadership was increasingly divided, in simplest terms into Independents and Presbyterians. Opposition to this regime resulted in petitions to parliament, most of them calling for a personal treaty with Charles, and the 'settling' of the kingdom; in riots, particularly against taxes and restrictive religious policies; and in personal opposition to imposed government, such as the unpopular county committees.[3] The maritime community, supported by overseas

---

*Paper presented at the fourteenth annual conference of the Naval Dockyards Society held at the National Maritime Museum, Greenwich on 17 April 2010. Theme: Pepys and Chips. Dockyards, Naval Administration and Warfare in the Seventeenth Century.

merchants, were also deeply concerned about the 'great sufferings by Pirates in the narrow seas' which had yet to be addressed.[4] Due to both Charles's intransigence and the extent of his opponents' demands, negotiations never came close to reaching a satisfactory compromise, but the continued hope for a personal treaty and a return to traditional government is palpable.[5] Robert Ashton describes this as 'a groundswell of enthusiasm for the king', provoked particularly by anxiety over the possibility of a kingless Britain, which intensified after the vote of no addresses, taken by parliament in January 1648, ended the negotiations between parliament and Charles.[6]

For many, the increasingly politicized army represented the greatest threat.[7] Not only were they thought to be the root cause of the hated excise and assessment taxes, they were held responsible for various breaches of law, and were popularly associated with the radical religious sects which many saw as an obstacle to the reestablishment of a unified national church.[8] This reaction was also provoked by the experience of having to billet soldiers, who often exacted free quarter when their pay was in arrears, and this the maritime community experienced at first hand. Sir Robert Pye's regiment was quartered in Deptford in the summer of 1647, when there were objections about their behaviour, and there were also complaints from Chatham workers that they had been 'forced to billet soldiers'.[9] Although Pye's was not in fact a new model regiment, the distinction between different kinds of soldier does not seem to have been generally acknowledged. Presbyterian attempts to disband the army without pay, and to employ the London militia as a countering force, served only to alienate the soldiery further, and in June 1647 the army seized the king, giving them considerable political leverage and a central role in negotiations.[10] When anti-army sentiment broke out in petitions, followed by violent rioting in London during July and August, both of which apparently involved seamen, the army occupied the capital.[11] Despite the king's escape to the Isle of Wight in November, and notwithstanding divisions due to the Leveller movement, the army and their political allies were in the ascendant at the beginning of 1648.[12]

For the maritime community and the Navy, this ascendancy had controversial consequences. From 1642, the naval administration and operational command were in the hands of the Earl of Warwick, a puritan peer with privateering interests, supported by men drawn from the mercantile and maritime elite of London.[13] This control of the Navy had given parliament and its supporters the most powerful instrument of sea power in the British Isles, although they did not have undisputed control at sea.[14] Piracy was rife, especially in Irish waters, and trade undoubtedly suffered from increased violence and lawlessness, while the capturing and recapturing of ports such as Newcastle and Bristol made it difficult to prevent trade with the king's supporters.[15] Nevertheless, while Charles issued commissions to privateers, he was unable to organise a unified naval force, which has often been considered a significant factor in his eventual defeat.[16] Moreover, both the naval command and the maritime community seem to have remained committed to the parliamentary cause throughout the first civil war.

Warwick was dismissed by the self-denying ordinance in 1645, by which all members of parliament were removed from military command, and Vice-Admiral William Batten was made commander-in-chief at sea.[17] Batten has generally been designated a Presbyterian, and though there is little actual evidence concerning his personal religious views, it is clear that he was hostile to the sectarianism of the Army and Independents.[18] In 1647, following the riots in London, the Army's political allies in parliament impeached eleven Presbyterian MPs, who received permission to go abroad to prepare their legal defence; they were stopped by a naval ship while on their way to France, but subsequently released by Batten.[19] Bernard Capp described this as an 'affront' which caused the Independents to review control of the Navy, though Batten himself seemed to think that the hostility ran deeper, writing later 'I was displaced by a Committee…because I was not of the temper of the *Army*.[20] Batten was called before the Admiralty committee, which had been expanded to include Independent members such as Colonel Thomas Rainsborough and Henry Marten, and resigned his commission, though he protested 'it was not out of any discontent', and he continued in the naval administration.[21]

Batten was replaced by Rainsborough, whose father William had been an important London shipmaster, and commander of a royal expedition against North African pirates in 1637.[22] Nevertheless, as a radical army officer, Thomas Rainsborough proved a contentious choice. *Mercurius pragmaticus*, a royalist newsbook and therefore an admittedly hostile source, called him 'one of the *Princes* of the LEVELLERS', and this is certainly the reputation he has been given among historians, being particularly well known for his part in the Putney debates of 1647.[23] He was only finally accepted by the Lords in April 1648, although he was at sea before then, without the Lords' approval, when Charles I escaped his army captors and fled to the Isle of Wight, only to be incarcerated again in Carisbrooke Castle.[24] Rainsborough was sent to secure the island, following 'strange' rumours circulating that Charles had attempted to escape, or that foreign intervention might liberate him.[25] In October 1647, also, the committees of the Navy and Army were ordered jointly to consider and secure the 'Habiliments and Materials of War' belonging to the army and navy.[26] Thus, by the early months of 1648, among widespread dissatisfaction and growing opposition to the parliamentary regime, the naval command had come under the influence of the Army and Independent faction, although Warwick, Batten, and others important during the first war continued to be involved in the naval administration and to hold commands at sea.

The Role of Dockyards in the Maritime Community

The royal dockyards of the Thames had achieved a new importance during the reign of Charles I, when naval activity had increased significantly with the 'ship money' fleets of the 1630s, and they continued to be crucial throughout the 1640s.[27] As part of this increase in activity, Charles established a new programme of shipbuilding, culminating most famously in the *Sovereign of the Seas* in 1637, which stimulated more work for the naval dockyards than had been seen during his father's time.[28] Although there was a lacuna during the conflict years of the early 1640s, in 1647 parliament resolved to supplement the fleet with four new frigates, to be built at Woolwich, Deptford, and Chatham.[29] This work was clearly intensive; Daniel Larkin, a shipwright at Deptford, was ordered to be excused parish duties because he was 'so necessary an instrum[en]t' that 'hee cannot possibly bee spared' from his work in the yards.[30]

On a deeper level than just building new ships, however, the dockyards and stores were an indispensable part of early modern naval infrastructure, due to the material demands of seventeenth century shipping. As maritime archaeologists have pointed out, ships were probably the most complex machines of the pre-industrial world, and moreover machines in frequent need of maintenance they could not receive at sea.[31] The physical fabric of a ship, as well as food and other supplies, were constantly subjected to the onslaughts of use and bad weather, and the Navy Board bill books give an eloquent testimony of the sheer mass of material required to keep the navy's ships afloat.[32] The importance, and cost, of this material is underscored by complaints from the Admiralty committee 'that divers stores of [th]e Navy are frequently imbezelled'.[33]

Ships were regularly 'sent up' to Chatham, Woolwich, or Deptford, for refitting and repairing, so that the naval dockyards were centres of continual activity, and the records of the Ordinance Office for the later 1640s show that Chatham and Deptford were by far the most active yards.[34] This intensified at certain times of the year because the fleet was not constantly at sea: the navy was sent out in 'expeditions' in the summer and winter, creating surges of activity in the spring and autumn.[35] Moreover, naval activity increased steadily throughout the civil wars. According to one estimate the cost of 'grounding graving, rigging caulking carving painting, sailes, grountacle, seastores, petty pr[o]visions &c' was £35,640 in January 1642; the eventual cost of the summer and winter guards of 19 royal and 23 merchant ships for that year was £204,810 16s 3d.[36] In 1647, 43 royal ships and 16 merchant vessels were set forth, at a total cost of £244,655; an increase of 17 vessels (but of 24 state-owned vessels), and costing approximately 19% more.[37]

There was an important political aspect to this activity in the dockyards, both before and during the civil wars. Naval dockyards and dockyard workers were part of the state apparatus – the 'network of intermediaries', from the Privy Council to parish constables, which Michael Braddick identifies as the early modern state.[38] As the Navy represented one of the major manifestations of the state with which the maritime community had contact, the royal dockyards were a focal point for their relationship with it, both as a part of the process through which the state's agency was carried to sea in the Navy, and as a place where they might themselves be involved in that process. The role of the Navy was articulated more clearly with Charles I's attempt to enforce his 'sovereignty of the sea', and place the waters surrounding Britain under his own *pace Domini Regis*, literally expanding the state beyond its geographic borders.[39] Perhaps paradoxically, parliament also attempted to enforce this sovereignty throughout the civil wars and even after Charles I's execution in 1649.[40] This state involvement meant that dockyards were potentially volatile centres for political debate and dispute, as they were to become in 1648.

As well as employing the Navy, and the naval infrastructure, to create and defend their own vision of the state, parliament also used the dockyards to further sometimes controversial religious reform. Mr Clare, the 'Minister of Gods word at Chatham' who had been appointed by parliament, reportedly caused contention and in March 1647 some 'Officers or other persons employed in the Navy' objected and tried to remove him, though the Admiralty committee wrote to him expressing their support, and ordered the names of the troublemakers to be sent to them.[41] This shows the depth of opposition among the maritime community to parliament's programme of religious reform, and again highlights how dockyards could become important centres of political conflict, as sites where the 'network of intermediaries' interpreted – or confronted – the decisions of central government.

Naval shipping and royal shipyards, however, did not exist in isolation, and private dockyards were also important spaces in the maritime community. Shipyards, especially those at Deptford and Woolwich, were part of the Thames-side parishes which were home to the largest maritime community in Britain.[42] According to John Stow, writing in 1603, the development of private dockyards had been a catalyst to the growth of these parishes from the middle of the sixteenth century, concomitant with the phenomenal growth of the shipping trade during this period.[43] Dockyards were a place of connection between the land and the sea, part of the unique, maritime nature of these parishes, producing and sustaining the material subculture which defined the physical experiences of seafarers.[44] They were part of the wider web of craftsmen, which included chart- and instrument-makers also based in these parishes, generating the maritime and navigational knowledge and equipment which made early modern seafaring possible.[45] This connection is highlighted by the office of a 'lecturer to the navy' maintained at the royal dockyards to teach navigation during the early seventeenth century, discontinued by parliament in 1641 but reinstated in 1649.[46] The dockyard at Chatham was also a place of departure, where pressed sailors, at least 793 in 1648 alone, were sent to be put aboard ship.[47]

Socially, too, dockyards functioned as focal points. The craftsmen who owned and were employed in them were part of the more stable, shore-based element of the maritime community which maintained cohesive, connecting relationships with the unstable, mobile and swiftly changing community of seafarers. Some indication of these relationships survives in the wills of mariners naming shipwrights and other maritime tradesmen as executors or overseers, positions of considerable trust and responsibility.[48] Maritime tradesmen also appeared as securities in indictments against mariners and others in the Thames parishes, and vice versa, again indicating associations of mutual responsibility.[49] Some of these tradesmen, such as shipcarpenter Abraham Sanford and shipwright Thomas Mabstow, both of Poplar, were appointed as collectors for the poor in Stepney parish.[50] This evidence survives only because they were in fact indicted for neglecting their duties, but it shows that maritime tradesmen were nominally expected to be involved in the community they inhabited.

Shore-based maritime trades were also more regulated than the comparatively open seafaring professions: a Brotherhood of Free Shipwrights had existed in London since medieval times, and a new corporation based at Rotherhithe was incorporated by James I, and received royal support in subsequent disputes between the two companies, being expanded to include other associated trades such as caulkers.[51] This attempt at increased government influence over these trades implies recognition of their significant role, especially in the maintenance of the Navy, but the disputes show that it was not entirely successful. During the civil wars, another dispute arose concerning the corporation's officers, who complained to the Admiralty committee because they were not being paid their wages.[52] This lack of support, and the fact that very few members of the corporation responded to the committee's summons to appear before them, suggests that the trade was somewhat divided, or at least that the new corporation had not established universally-acknowledged authority.[53]

Though not all maritime tradesmen accepted the state-supported corporation, it maintained influence through its close links with the naval dockyards during this period. When planning the building of the new frigates early in 1647, the Navy Board contacted the Shipwrights' Hall and Trinity House, the corporation of shipowners of the Thames, requesting '3 or 4 of the ablest shippwr[igh]ts of yo[u]r Hall' along with '2 or 4 of the ablest Elder Brethren [of Trinity House]' to be sent to the naval dockyards.[54]

Geoffrey Scammell has argued that shipbuilding on the Thames, for much of the early modern period, was 'in the hands of a small group of leading shipwrights, many of whom also worked for the Crown'.[55] Of these, the Pett family were the most influential, a dynasty of shipwrights including Phineas Pett, master shipwright at Chatham, who was the first master of the new company of shipwrights, and a governor of the Chatham chest.[56] He and his son Peter designed and built the *Sovereign*, and Peter was a member of parliament's naval administration, an owner of private Thames shipyards, and a churchwarden of Stepney parish in 1647 and 1648.[57] Peter also built, in his private yard, the *Constant Warwick*, whose part-owners included Warwick, William Batten, and Robert Moulton and Richard Swanley, two London shipmasters who served as parliament's vice-admirals for the Irish seas, indicating again the close connections between the navy and the maritime community.[58] Pett was ordered in February 1647 to build one of the new frigates at Deptford, and his younger brother Christopher, who had been lieutenant of the *Garland*, was thereafter given a place as master shipwright's assistant at Woolwich.[59]

It is unsurprising, then, that control of the Thames dockyards became an important issue during the conflicts of 1648. On a purely practical level, the dockyards were fundamental to setting forth the navy, not only in meeting the government's demands for more ships, but also in preparing and maintaining those already in existence. This connection with the navy, as the military expression of the state at sea, meant also that the dockyards and the craftsmen employed there were involved in the ongoing process of state formation, even if they were occasionally individually opposed to the wishes of the government. More generally, too, dockyards were focal spaces in the community, their owners significant and influential figures, their workers deeply connected with the maritime culture and identity.

The Contest for Control: May 1648

The second civil war in 1648 was a series

of uncoordinated rebellions against parliamentary rule, sharing general aims but not a strategy. Charles I signed an Engagement with his Scottish supporters to raise an army and renew the war in December 1647, but the first actual uprising was the mutiny of parliamentary troops in Pembroke in March 1648.[60] Early in May, a group of gentlemen in Kent began to circulate a petition calling once more for a personal treaty with the king, and when parliament's response was uncompromising, the petitioners began to gather forces, including former royalist soldiers.[61] This provoked alarm concerning the fleet stationed in the Downs, off the Kent coast, and the dockyards on the Thames, but the papers of the Admiralty committee show little awareness that there might have been disaffection in the navy, though Henry Oxinden, an apparently unaffiliated Kent gentleman, noted rumours of mutiny early in May.[62] Rather, the Admiralty committee were afraid that the stores at Chatham and elsewhere might be plundered, or that artillery might be used to threaten shipping in the Thames estuary.[63]

These fears were soon realised. On 23 May Peter Pett, then in command at Chatham, mustered the dock workers and shipkeepers, and found some 'missing which were joyned in that horrid engagement'.[64] The same day the Kent petitioners sent a messenger to Chatham, asking Pett to sign the petition, to let them circulate it in the yards, and to allow them to commandeer ordinance, all of which he refused. They repeated this last request three days later, which Pett again refused, though he complained he was 'forsaken almost by the whole Navy', and on the same day a report reached the commissioners of the Navy that 'some people are now at Deptford, seizing upon & plundering the Stores there', including some small vessels intended for Guernsey.[65] The disaffection of the dockyard workers was presumably influenced by their pay being heavily in arrears, and their petitions for payment having been ignored, for a later report stated that they had received no pay since 1646, though it would be unfair to assume that only material concerns influenced their opinions.[66] Rainsborough was ordered to send reinforcements to Chatham, and Robert Moulton and Edward Hall, London shipmasters who had both previously served in the Navy, were ordered to take command of the *Sovereign* and the *Prince*, lying in the Medway.[67] An appeal was also made to Trinity House, that all 'well affected persons...vnite their best endeuo[u]rs' to protect shipping.[68]

The Kent commanders sent more envoys to Pett on 27 May, which he again rebuffed, but the day after a company of soldiers arrived at Chatham, 'ushered by some of the principall gunners of the navy to the new dock', and Pett was forced to bar the gates and protest 'out of a window'.[69] Though he protected the dockyards and stores, the *Sovereign* and *Prince* were carried away: it is not clear whether Moulton and Hall ever arrived, but it would seem that Trinity House, and the majority of Thames ship-masters, had ignored the Admiralty committee's requests. The encounter at Chatham was also a crucial element in the process by which affairs in Kent escalated from petition to revolt; on 23 May, the messengers denied 'perfidiousness to the Parliament', but five days later they protested that parliament 'had proclaymed them rebells and traytours, and they were resolved to defend themselves as long as they could'.[70] The successes at the dockyards gave the uprising confidence: one newsbook reported that by

> *securing Rochester, and the Magazines at Chattam and other places, they have collected themselves, by the assistance of Sea-men, Watermen, &c into such a posture, that they have secured the Country.*[71]

While this might be exaggerated, it is significant that parliament's only response was to offer indemnity should they surrender what they had taken, and depart peacefully, adding that only then would any petitions be considered.[72]

The process of escalation was influenced also by the mutiny of a number of the fleet in the Downs. The first warning that the crews of the naval ships might share the sympathies of the Kent forces came in a letter to the House of Lords from Rainsborough, who then went ashore on 27 May to secure important coastal castles.⁷³ The same day Samuel Kem, the parson of Deal, a former parliamentarian officer who had also been Batten's chaplain, and who was associated with the Kent gentlemen, went to the fleet to circulate the Kent petition and encourage support from the sailors.⁷⁴ When Rainsborough returned to his ship, the *Constant Reformation*, he was told by the crew that 'they would obey him no longer, but would have the King brought to London', and that coming aboard would be 'at his perill'.⁷⁵ The sailors of the *Reformation* then called a council, to which officers of other ships came.⁷⁶ Captain Francis Penrose objected and, when armed men came to summon him in the name of the vice-admiral, asked who this vice-admiral was; upon being told '*his name was* Lendall', Penrose replied '*he knew such a man to be a Bostons* [i.e. boatswain's] *Mate, but no otherwise*'.⁷⁷ It is intriguing that a low-ranking officer was named as the initial leader, and a later list of officers held guilty by parliament included a number of pursers, gunners and boatswains, suggesting that this was not a decision taken by the commanding officers alone.⁷⁸ Despite his reluctance, Penrose attended the council, apparently forced by his own crew, but once there refused to sign the petition or acknowledge Lendall, suggesting that they 'make choice of some honourable person, that is true to both [i.e. king and parliament]'; at which, reportedly by unanimous consent, Warwick was selected.⁷⁹

Penrose was dispatched with letters to Warwick and to the commissioners of the Navy, the latter printed soon after as *The declaration of the navie*, signed by nine officers of the *Reformation* and the *Swallow* – Lendall is curiously absent from these signatories, throwing into doubt the actual significance of the role he played.⁸⁰ The sailors proclaimed themselves 'unanimously joyned with the *Kentish* gentlemen', and called for a personal treaty with Charles, along with the paying and disbanding of the Army, and that 'the known Laws of the Kingdome may be Established and continued…[and] That the Priviledges of Parliament and the Liberty of the Subject may be preserved'. These demands show that the mariners did not consider themselves irrevocably separated from the parliamentary leadership, although they reveal dissatisfaction with the regime, and are similar to demands in other contemporary petitions.⁸¹ Another version of the *declaration* printed the oath taken by the mariners to

> *endeavour to maintaine the Glory of God, the purity of that Religion which is most agreeable to the Word of God, the Honour, Freedom, and Preservation of His Majesty, the Priviledge of Parliament, and the Liberty of the Subject.*⁸²

This demonstrates how the persuasive traditional model of loyalty to both king and parliament maintained its enduring appeal even after the first civil war, and many clearly still interpreted political events according to this model.

There is little in the *declaration* concerning Rainsborough himself, except that the sailors 'resolved to take in no Commander whatsoever, but such as shall agree and correspond with us in this Petition'.⁸³ This undermines the prominence most historians have accorded Rainsborough in causing the mutiny; it appears that he was removed because it was known he would not 'agree and correspond', rather than as a reaction against his own behaviour, an interpretation also supported by the account of one of the Kent petitioners.⁸⁴ Even so, his '[i]gnorance and insolency', described in an un-attributed manuscript note, may have provoked feeling against him, and there appears to have been serious dislike of the Army's intrusion into maritime affairs, as other army officers had also been offered naval commands.⁸⁵ Yet this was not only a reaction against the Army, as the demands of the *declaration* reveal a deep concern for the more general issues at stake, and a desire to influence central government policies.

The *declaration* shows, like the encounter at Chatham, that these were not initial moves towards open conflict; the addresses to parliament were couched in respectful terms, and the aim of the petitioners was to persuade, not overthrow, the government. However, because of the importance of dockyards as political and military spaces, the disorder at Chatham and Deptford, coupled with the mutiny of the sailors, contributed to the alarm with which parliament viewed these developments and thus intensified the escalation from petition to rebellion, although this rebellion did not last long. Parliamentary forces arrived under Lord General Fairfax, and dispersed a gathering, reported to include mariners, on Blackheath outside London at the end of May; soon after Fairfax took Maidstone, despite bitter resistance from the Kent forces.[86] Early in June the last of the royalist army in the southeast were besieged in Colchester, a siege which lasted until the end of August.[87] The immediate military threat to the dockyards was removed, and some of those who had earlier deserted their posts now returned, although Pett ordered the dockyard clerk to disallow them 'victualls and wages till such tyme as they could cleere themselves'.[88] In June, parliament ordered a letter of thanks to Pett, for his '[d]eportment during the late Rebellion in Kent, for the Safety of Chatham Yard, [and] the Ships and Stores there'.[89] Nevertheless, the mutiny of the fleet had raised serious questions about the loyalty of the maritime community, and this was to be the subject of a second struggle, lasting throughout the summer of 1648.

The Contest for Loyalty: June-November 1648

In early June, while the situation on land was more stable and under parliament's control, this was not the case at sea. Parliament named Warwick admiral as the sailors of the fleet requested, but when he came to the Downs they would not accept him unless he supported the Kent petition, which he refused to do.[90] Following this, he hastened to Portsmouth, from where he wrote to parliament that the companies of ships there 'have severally ingaged themselves to live, and dye with mee in the Parlyaments Cause'.[91] However, a letter written by two navy commissioners with him, Batten and Richard Crandley, stated 'we find [th]e Seamen disaffected and poisoned…there is not a possibility of quieting the seamen unless some suddaine addresses be made to the king', and indeed the *Anthelop* rejected Warwick's overtures and sailed to the Downs to join the fleet.[92] This second letter probably represents another attempt to persuade the parliament, and throws some doubt upon Batten's later claim that he had transferred his loyalty to the Prince of Wales by this point.[93] Indeed, it suggests that for Batten, loyalty to both parliament and king were still not incompatible, and that he was using his position within the naval administration to further the cause of a peaceful resolution and a personal treaty.

These goals, however, were becoming ever more distant. Following the defeat of royalist land forces, on 12 June the fleet sailed for Holland, to be received by the Duke of York, later being joined by the Prince of Wales; meanwhile, one report declared '[t]he falling off of those ships…makes other Sea-men tumultuous'.[94] Faced with this spreading hostility, parliament voted on 17 June to raise a new fleet, and Warwick, now returned to London, wrote to the commissioners of the navy ordering that

a perfect disquisition [be] made, whether any, and what Officers of the Navy, or Yards, at Deptford, or Woolwich were in any, and what kinde, ingaged in the late Rebellion in Kent

while Pett conducted investigations in Chatham.[95] A list of those 'active against the parliament', compiled two days later, included the master joiner, cooper, and pumpmaker, as well as a number of ships' officers, and Thomas Granee, 'Minister of the Navy'.[96]

Warwick had also written to Trinity House for assistance in preparing the new fleet, and their response is revealing as to feeling in the wider maritime community.[97] At a meeting on 21 June, to which both members and non-members of the corporation were invited, it was decided, instead of responding to Warwick's letter, to petition parliament for a personal treaty with the king, and a peaceful resolution of the mutiny.[98] Not long afterwards this petition was printed, along with others presented to Trinity House and to parliament, in a pamphlet which showed the strength of feeling in the maritime community against renewed conflict, and on behalf of the traditional government and social order.[99] These petitions appealed to the ideal of a communal identity, for

> *it cannot be imagined that the Seamen of England, who are as it were in a Fraternity will be drawn to fight one against another, since both pretend and desire one and the same thing: The desire of a Personall Treaty…is earnestly desired by the generality of all Seamen*[100]

These claims to represent such a 'Fraternity' were soon proved to be false; a counter-petition supporting the raising of a new fleet was presented on 5 July, begun by those who had initially opposed the 21 June petition at the Trinity House meeting, and led by Robert Moulton.[101] These were a minority, only 51 signatures compared with 468 to the Trinity House petitions, but not as socially insignificant as royalist newsbooks were swift to portray them.[102] One royalist writer claimed that some of the 5 July petitioners were '*Sayle-makers* by their trade', and while this was an attempt to cast aspersion on their reputation, it suggests that this division was not restricted to seafarers alone.[103] It is therefore apparent that a proportion, albeit a small one, of the maritime community remained committed enough to parliament's political and religious cause to feel that more conflict was justifiable.

The situation remained tense throughout July, as Warwick attempted to prepare a fleet with which to face the prince, who returned with his ships to the English coast early that month. According to royalist reports, Warwick had to recruit Newcastle coal ships to supplement his numbers, and even so by the beginning of August had no more than 'six inconsiderable ships'.[104] The maritime community remained deeply divided: those who supported Warwick later complained 'we receive many *reproaches*, nay, *assaults* and *affronts*, not only to the detriment of our *names*, but the hazard of our *lives*'.[105] The sailors pressed into the fleet were disorderly and reportedly told Warwick that their consciences forbade them from fighting against the prince; in response, Warwick imposed martial law, a move which proved even more unpopular.[106] At the same time, there was anxiety over the questionable loyalty of the dockyard workers: investigations were carried out by Warwick himself, who wrote in mid-July

> *I had examined many of them and absolved some, the evidence against them failing, but if the House had anything against them…they must submit to further examination. As some of them…have been of great use to me in fitting out the ships, and as their going up [to prison] will be a great hindrance I have ventured to keep them till I be got out.*[107]

Warwick's comments again reveal how important the dockyards of the Thames and their workers were to naval activity, and his 'getting out' was urgent, as the prince's fleet, largely unopposed, interrupted trade and shipping, as well as enabling a few limited and ultimately unsuccessful excursions on land.[108] It may have been the prince's own actions which subdued royalist sentiments in the maritime community, for while they had been outspoken for a personal treaty with the king and a peaceful resolution, there was little tangible support for the prince's military cause. In July William Batten joined the disaffected fleet, with another important London shipmaster, Elias Jordan, who had been involved in the 21 June Trinity House petitions, taking with them the *Constant Warwick*.[109] In August William Hawkeridge, formerly a parliamentarian privateer, sent a small ship to join the prince, but this was intercepted in the Thames.[110] These appear to be isolated incidents, and despite the apparent unpopularity of Warwick's martial law, there is certainly no evidence of a large movement to join the prince, in spite of declarations from the sailors of the royal fleet encouraging mariners to do so.[111]

August witnessed more explicit attempts to justify each side of the division in the maritime community, written for the royalists by Batten, and for the parliament by Richard Badiley, a shipowner and master influential in Mediterranean trade, who had been involved in the 5 July counter-petition.[112] These arguments centred on the interpretation of the Solemn League and Covenant, an oath taken by parliament's supporters from 1643 onwards. While these publications were explicitly aimed at seafarers, the majority of the London maritime tradesmen would also have taken the Covenant, London having been in the parliamentary heartland throughout the war; this debate would thus have had significance for them as well. Indeed, the Admiralty committee had ordered a check in April 1647 on who had and had not taken the Covenant at Chatham, showing its importance as a marker of loyalty to parliament.[113]

Batten claimed that 'the Solemn League and Covenant binde[s] us to the preservation of Religion, and Liberties, and to maintaine, and defend the Kings Person and Authority', all of which the parliament had betrayed.[114] It was therefore entirely consistent with the oath to return their loyalty to Charles I, and to serve in the fleet commanded by his son: indeed, Batten claimed 'I believe I have kept [the Covenant] better in leaving the Parliament...for they have put downe both Religion, and the King'.[115] Badiley, by contrast, argued that keeping the Covenant meant loyalty to parliament's cause, and appealed to the maritime community by stating that since parliament had sat, 'men of your own coat have been placed at the Stern to manage the maritime affairs of the Kingdom'.[116] He compared the mariners' mutiny to the betrayal of Brutus, emphasising that they had been in a position of public trust, which they had abandoned out of personal motives, for 'I cannot but discerne that many amongst them have been in the service of King and Parliament...and thereby gave [the King] such great distast'.[117] This question of trust and loyalty, and to whom it was due, would also have been particularly relevant to those in the royal dockyards; it was a question whether they were to continue to serve the parliamentary state, as it was fast becoming, or return to their traditional loyalty to the king. Even after the first civil war, as the petitions of May and June 1648 show, the principle of loyalty to both parliament and the king remained intact, if not undamaged, but it was no longer tenable following the renewed conflict of the second civil war, and the situation of emergency created by the presence of the prince's fleet.

It would seem that for the majority of the maritime community, neither of these extremes was particularly favourable; on the one side, there was no outpouring of support for the king, but on the other Warwick's command, and especially the imposition of martial law, remained unpopular throughout August.[118] This widespread reluctance to support further conflict, in the end, favoured parliament's stronger position, as each successive royalist uprising failed, culminating in the defeat of the Scottish Engager army at Preston on 17-19 August, and the surrender of Colchester not long after.[119] Despite his difficulties, Warwick was able to gather and man a substantial fleet by the end of August, due in no small part to control of the dockyards.[120] Without the resources provided by the dockyards, it is unlikely Warwick could have prepared ships for sea, and it is probable, too, that control of the dockyards had symbolic significance, giving credibility (among the maritime community at least) to parliament's claims to control the organs of state. Just as the defeat of the other uprisings and the Scottish army weakened royalist prospects, parliament's control of these focal areas may have dissuaded the maritime community from open support for the prince, even if they had no committed loyalty to parliament. The activity of influential figures in the parliamentary causes, such as Badiley, Moulton and Pett, presumably also had a significant impact on the opinions of maritime tradesmen and the community as a whole.

Warwick maintained control of the Thames and, when the Prince's fleet advanced into the estuary at the end of August, Warwick was able to meet them, although, prevented by a storm, the fleets did not engage one another.[121] Faced by Warwick's ships, Prince Charles chose to retreat. Warwick pursued him to Hellevoetsluis in Holland, where he blockaded the prince's fleet until November, when a number of the prince's mariners, their morale weakened by infighting amongst the royalist leadership, submitted to Warwick and accepted his indemnity; among them were Batten and Elias Jordan.[122] Though a few ships remained under the command of the new royalist admiral Prince Rupert, who set to sea following Warwick's departure in late November, they were quickly bottled up in Kinsale.[123] The naval revolt, and indeed the second civil war, ended once again in military victory for the new model army and its allies, which on 6 December famously purged parliament.[124]

Conclusion

In January 1649, an act of parliament ordered that all customs officials, dockyard workers, naval officers, and Elder Brothers of Trinity House who had been involved in the royalist cause from 1641 onwards, or had supported a personal treaty or the 1648 revolt, were to be removed.[125] This purge, which placed effective control of the Navy and London shipping in the hands of those loyal to parliament and in political affinity with the Army (an affinity made explicit by a declaration printed in December), was necessitated by the uncertainty of the loyalty of the maritime community, revealed by the naval revolt and the defection of dockyard officials to the royalist cause during the second civil war.[126] The need to protect the dockyards and stores encouraged parliament to resort to force, and by alarming them into a hostile response, intensified the escalation of the May and June petitions into open conflict between those still committed to the parliamentary cause, and those who felt that matters had gone too far, and should be resolved by a personal treaty restoring Charles I to power. Even after parliament had re-established firm control of the dockyards, the struggle continued, becoming a contest for the loyalty of dockyard workers, and the maritime community as a whole: in this contest, control of the dockyards proved important to parliament, allowing Warwick to prepare a fleet with which to face Prince Charles, and lending credence to their claims to control the state. Although the disputes of the summer of 1648 may not have persuaded many in the maritime community to take action, the very debate itself polarised the arguments involved, making a peaceful resolution ever less likely. These disputes, the result of the attempt by supporters of Charles I to pursue a personal treaty, first by petition and then by force, contributed (perhaps with tragic irony) to the image of Charles as a 'man of blood' with whom no compromise could be reached, and led ultimately, though not inevitably, to his execution.[127]

Acknowledgements

For reading drafts of this paper and submitting comments, I am grateful to my supervisor, Dr David Smith, and to Dr Alan James.

References

1. BL, Add. MS 9300, fo. 91r.

2. Robert Ashton, *Counter-revolution: the Second Civil War and its Origins, 1646-1648*, Yale University Press, 1994; see also John Morrill, *Revolt in the Provinces: the People of England and the Tragedies of War, 1630-1648*, Longman, London, 2nd edn, 1998, 169-76; Ian Gentles, *The English Revolution and the Wars in the Three Kingdoms, 1638-1652*, Pearson Longman, Harlow, 2007, 291-322; Michael Braddick, *God's Fury, England's Fire: a New History of the English Civil Wars*, Allen Lane, London, 2008, 465-506.

3. Ashton, *Counter-revolution*, 109-138; Martyn Bennett, *The Civil Wars in Britain & Ireland, 1638-1651*, Blackwell, Oxford, 1997, 295-6; Michael Braddick, 'Popular politics and public policy: the excise riot at Smithfield in February 1647 and its aftermath', *Historical Journal*, 34, 1991, 597-626; for another riot in London see *Moderate Intelligencer*, 160, 6-13 April, 1648, 12 [1269]. For all newsbooks, the first number is the page within the individual issue, the number in square brackets is the page or sig. number within the whole series. Place of publication for primary sources, unless otherwise stated, was London.

4. Quoting *Moderate Intelligencer*, 158, 23-30 March, 1648, 2 [1234], reporting a petition to parliament from merchants; for complaints concerning piracy, see *Journal of the House of Commons*, v, 130, 131, 247, 505; *Journal of the House of Lords*, ix, 337; *Pefect Diurnall of some Passages in Parliament* [hereafter *PD*] 192, 29 March-5 April 1647, 5-6 [1539-40]; BL, Add. MS 9305, fo. 17r-18r, 26r, 50r; Add. MS 9306, fo. 98v; TNA, ADM 7/673, 127.

5. On Charles's questionable commitment to negotiations, see Austin Woolrych, *Britain in Revolution, 1625-1660*, Oxford University Press, 2002, 408; Gentles, *The English Revolution*, 291, 296, 314; Braddick, *God's Fury, England's Fire*, 467-73.

6. Ashton, *Counter-revolution*, 205ff.

7. For the politicization of the army, see Mark A Kishlansky, *The Rise of the New Model Army*, Cambridge University Press, 1979, esp. chs. 7-8; Ian Gentles, *The New Model Army in England, Ireland and Scotland, 1645-1653*, Blackwell, Oxford, 1992, 139-41.

8. Ashton, *Counter-revolution*, 159-65; Bennett, *The Civil Wars*, 266-7.

9. *Journal of the House of Commons*, v, 215, 263; TNA, ADM 18/3, fo. 38r.

10. For succinct accounts of these events, see Valerie Pearl, 'London's counter-revolution', in G E Aylmer, ed., *The interregnum: the Quest for Settlement, 1646-1660*, Macmillan, London, 1972, 29-56; Ian Gentles, 'The struggle for London in the second civil war', *Historical Journal*, 26, 1983, 277-305; idem, *The New Model Army*, chs. 6-7; idem, *The English Revolution*, 307-14; Braddick, *God's Fury, England's Fire*, 488-503.

11. *The humble petition of the citizens, commanders, officers, and souldiers of the trained bands and auxiliaries, the young men and apprentices of the cities of London and Westminster, sea commanders, seamen, and watermen*, [21 July] 1647; *Journal of the House of Commons*, v, 254-5, 268; *Journal of the House of Lords*, ix, 353-4, 357; London Metropolitan Archives, MJ/SR/1000, fo. 200r; MJ/SR/1303, fo. 101r. For all publications, dates given in square brackets represent the dating given in the Thomason Tracts.

12. For a brief introduction to the Leveller movement, and some of the writings of its key figures, see Andrew Sharp, ed., *The English Levellers*, Cambridge University Press, 1998.

13 Donald Kennedy, 'Naval captains at the outbreak of the English civil war', *The Mariner's Mirror*, 46, 1960, 108-42, and idem, 'The establishment and settlement of Parliament's admiralty, 1642-1648', *The Mariner's Mirror*, 48, 1962, 276-91; John R Powell, *The Navy in the English Civil War*, Hamden, Connecticut, 1962, ch. 1; Kenneth R Andrews, *Ships, Money & Politics: Seafaring and Naval Enterprise in the Reign of Charles I*, Cambridge University Press, 1991, 184-6. On Warwick's religious and political role, see J S A Adamson, *The Noble Revolt: the Overthrow of Charles I*, Weidenfeld & Nicolson, London, 2007, and Sean Kelsey, 'Rich, Robert, second earl of Warwick (1587-1658), colonial promoter and naval officer', *Oxford Dictionary of National Biolgraphy*.

14 Powell, *The Navy in the English Civil War*; N A M Rodger, *The Safeguard of the Sea: a Naval History of Britain 660-1649*, Harper Collins, London, 1997, ch. 28; Bernard Capp, 'Naval operations', in J P Kenyon and J H Ohlmeyer, eds., *The Civil Wars: a Military History of England, Scotland & Ireland 1638-1660*, Oxford University Press, 1998, 156-91. See also M L Baumber, 'Parliamentary and naval politics, 1641-1649', *The Mariner's Mirror*, 82, 1996, 398-408.

15 On piracy in this period, see M L Baumber, 'The navy and the civil war in Ireland, 1641-1643', *The Mariner's Mirror*, 57, 1971, 385-97, and idem, 'The navy and the civil war in Ireland, 1643-1646', *The Mariner's Mirror*, 75, 1989, 255-68; J Ohlmeyer, 'Irish privateers during the civil war, 1642-1650', *The Mariner's Mirror*, 76, 1990, 119-33.

16 Bernard Capp, *Cromwell's Navy: the Fleet and the English Revolution, 1648-1660*, Clarendon Press, Oxford, 1989, 2-3, and idem, 'Naval operations', 156, 176; Gentles, *The English Revolution*, 96; for a more critical appraisal, see Malcolm Wanklyn and Frank Jones, *A Military History of the English Civil War, 1642-1646: Strategy and Tactics*, Pearson Longman, Harlow, 2005, 12-13.

17 On Batten, see C S Knighton, 'Batten, Sir William (1600/01-1667), naval officer', *Oxford Dictionary of National Biography*.

18 Gardiner thought Batten a presbyterian: Samuel Gardiner, *History of the Great Civil War*, 4 volumes, Longmans, London, 1893, IV, 314; also Robert Ashton, *The English Civil War: Conservatism and Revolution 1603-1649*, Weidenfeld & Nicolson, London, 1978, 325. For Batten's own views, see *Journal of the House of Lords*, IX, 433; W[illiam] B[atten], *The sea-mans diall, or, the mariners card: directing unto the safe port of Christian obedience*, [17 August] 1648, esp. p. 8.

19 For an account of this favourable to Batten, see anon, *A declaration of the representations of the officers of the navy. Concerning the impeached members of parliament, transported beyond seas*, 1647.

20 Capp, *Cromwell's Navy*, 16; William Batten, *A declaration of Sir William Batten, late vice-admiral for the parliament, concerning his departure from London, to his highnesse the Prince of Wales*, [21 August] 1648, 2. In all quotations, italics are as in original unless otherwise noted.

21 For the record of Batten's dismissal, see TNA, ADM 7/673, pp. 379, 381; on Batten as a navy commissioner, see ADM 7/673, p. 408; ADM 18/3, fo. 36r; BL, Add. MS 9305, fo. 41v.

22 Brian Quintrell, 'Rainborow, William (bap. 1587, d. 1642), naval officer', in *Oxford Dictionary of National Biography*; see also W R Chaplin, 'William Rainsborough (1587-1642) and his associates of the Trinity House', *The Mariner's Mirror*, 31, 1945, 178-97; Andrews, *Ships, Money & Politics*, ch. 7. Various spellings of the name were used by contemporaries: 'Rainsborough' is preferred throughout.

23 M[ercurius] P[ragmaticus], 14, 14-21 December 1647, 5 [sig. O3r]. On Rainsborough, see Ian Gentles, 'Rainborowe [Rainborow], Thomas (d. 1648),parliamentary army officer and Leveller', in *Oxford Dictionary of National Biography*. For the Putney debates, see Michael J Mendle, *The Putney Debates of 1647: the Army, the Levellers, and the English State*, Cambridge University Press, 2001.

24 *Journal of the House of Commons*, v, 378, 406, 413; *Journal of the House of Lords*, IX, 459, 606, 615-6; Idem, x, p. 115.

25 *Journal of the House of Commons*, v, 356, 359; TNA, ADM 7/673, pp. 440, 478-9; *PD*, 230, 20-27 December 1647, p. 5 [1853]; Rainsborough's presence at the Isle of Wight occasioned comment from both parliamentary and royalist writers: *MI*, 147, 6-13 January, 1647[/8], 14 [1124]; *PD*, 232, 10-17 January, 1647[/8], p. 6 [1678]; *MP*, 19, 18-25 January, 1647[/8], pp. 5-6 [sig. T3r-v].

26 *Journal of the House of Commons*, v, 339.

27 On the 'ship money fleets', see Andrews, *Ships, Money & Politics*, ch. 6; Rodger, *Safeguard of the Seas*, chs. 26-7; Andrew Thrush, 'Naval finance and the origins and development of ship money', in Mark Charles Fissel, ed., *War and Government in Britain, 1598-1650*, Manchester University Press, 1991, 133-62.

28 See Brian Quintrell, 'Charles I and his navy in the 1630s', *The Seventeenth Century*, 3, 1988, 159-79, and Dr Ann Coats's contribution to this volume.

29 *Journal of the House of Lords*, VIII, 646-7; on the term 'frigate' see Andrew Thrush, 'In pursuit of the frigate, 1603-40', *Historical Research*, 64, 1991, 29-45.

30 BL, Add. MS 9306, fo. 89v.

31 Keith Muckelroy, *Maritime Archaeology*, Cambridge University Press, 1978, 3; Jonathan Adams, 'Ships and boats as archaeological source material', *World Archaeology*, 32, 2001, 292-310, esp. 300-3.

32 For the 1640s, TNA, ADM 18/1-5.

33 BL, Add. MS 9306, fo. 113r.

34 National Maritime Museum, CAD/C/5.

35 E.g. BL, Add. MS 9300, fo. 79r; Add. MS 63,788 B, fo. 26r; TNA, ADM 7/673, p. 169; as an example of these surges of activity, see the orders from the admiralty committee to the navy committee for setting out the 'Som[mer]s Expedition', February 1647[/8]: SP 16/518, fos. 10r, 12r.

36 BL, Add. MS 9300, fo. 62r; Add. MS 17,503, fos. 3v-4r.

37 *Ibid*, fos. 13v-4r. Giles Grene, one of the navy committee, gave the numbers as twenty royal and twenty-three merchant ships costing £201,761 in 1642, and forty-three royal and thirteen merchant ships costing £213,415 in 1647: Giles Grene, *A Declaration in Vindication of the Honour of the Parliament*, 1647, 9-10.

38 Michael Braddick, *State Formation in Early Modern England, c. 1550-1700*, Cambridge University Press, 2000, 27-36, 94.

39 This Latin term (lit. 'peace of the Lord King') is used in the instructions of Charles's lord high admiral, the earl of Northumberland: NMM, LEC/5, fo. 4v. On navies and state-formation, see Jan Glete, *Navies and Nations: Warships, Navies and State Building in Europe and America, 1500-1850*, Almqvist & Wiksell International, Stockholm, 1993.

40 On a dispute over sovereignty with Swedish ships, see Historical Manuscripts Commission, *Thirteenth Report, appendix, part I: the manuscripts of his grace the duke of Portland, preserved at Welbeck Abbey*, 10 volumes, Eyre and Spottiswoode, London, 1891-1931, [hereafter *Portland MSS*], I, 437; *Journal of the House of Commons*, v, 170; *Journal of the House of Lords*, IX, 178; *PD*, 197, 3-10 May, 1647, 5 [5679]; *MI*, 113, 6-13 May, 1647, 1-2 [1057-8]. Selden's defence of Charles I's sovereignty, *Mare clausum*, originally published in Latin in 1635, was printed in English during the Interregnum: John Selden (trans. Marchamont Nedham), *Of the Dominion, or, Ownership of the Sea*, 1652.

41 TNA, ADM 7/673, 243, 264.

42 On the size and distribution of the London maritime community, see Kennedy, *Ships, Money & Politics*, 221-4.

43 John Stow (ed. Charles Lethbridge Kingsford), *A Survey of London: reprinted from the text of 1603*, 2 volumes, Clarendon Press, Oxford, 1971, II, 70-2; on the growth of shipping, see Andrews, *Ships, Money & Politics*, ch. 1; Ralph Davis, *A Commercial revolution: English Overseas Trade in the Seventeenth and Eighteenth Centuries*, David & Charles, London, 1967, and idem, *The Rise of the English Shipping Industry in the Seventeenth and Eighteenth Centuries*, Newton Abbot, 1971.

44 Cheryl Fury has argued that there was an early modern maritime 'subculture', existing within wider European tropes: Cheryl Fury, *Tides in the Affairs of Men: the Social History of Elizabethan Seamen, 1580-1603*, Greenwood Press, London, 2003, 87-9.

45 On these craftsmen, see E G R Taylor, *Mathematical Practitioners of Tudor and Stuart Britain*, Cambridge University Press, 1954; Norman J W Thrower, ed., *The Compleat Plattmaker: Essays on Chart, Map and Globe Making in England in the Seventeenth Century*, University of California Press, London 1978; Gloria Clifton, *Directory of British Scientific Instrument Makers, 1550-1851*, Zwemmer, London, 1995. On navigation more generally in this period the standard work remains David W Waters, *The Art of Navigation in Elizabethan and Early Stuart England*, National Maritime Museum, London, 1958.

46 TNA, ADM 82/128, 17, 19; see also ADM 18/3, fo. 36r.

47 TNA, ADM 18/5, 229, 237, 256, 269; see also BL, Add MS 9304, fo. 11r.

48 Examples from the earlier reign of Charles I include: TNA, PROB 11/149, Proved 22 May 1626, 11/151, 26 May 1627, 11/154, 4 December 1628, 11/157, 31 May 1630, 11/187, 20 August 1641, 11/190, 30 March 1642.

49 For 1647-8: London Metropolitan Archives, MJ/SR/0995, fos. 55r, 61r; MJ/SR/0998, fo. 110r; MJ/SR/ 0999, fo. 102r; MJ/SR/1002, fo. 13r; MJ/SR/1003, fos. 65r, 114r, 115r, 116r, 117r, 123r, 124r; MJ/SR/1007, fo. 36r; MJ/SR/1009, fo. 104r; MJ/SR/1011, fo. 59r, 100r; MJ/SR/1014, fos. 99r, 133r; MJ/SR/1015, fo. 91r; MJ/SR/1018, fo. 123r.

50 London Metropolitan Archives, MJ/SR/0990, fos. 106, 112.

51 On the corporation, see Phineas Pett (ed W G Perrin), *The Autobiography of Phineas Pett*, Navy Records Society, London, 1918, pp. xv-xli, at xxxiii. The 1605 and 1612 charters are printed in *ibid*, Appendixes III and IV, 176-206. On the disputes with the old company and the caulkers, see *ibid*, pp. xxxiv-xxxvi; TNA, SP 16/264, fo. 145r; SP 16/353, fos. 19v, 41v-2r, 87r-v.

52 TNA, ADM 7/673, p. 101.

53 TNA, ADM 7/673, pp. 119, 130.

54 *Ibid*, fo. 89r; cf. *MI*, 97, 7-14 January, 1647, 6 [488]. On Trinity House, see G G Harris, *The Trinity House of Deptford, 1514-1660*, Athlone Press, London, 1969.

55 Geoffrey Scammell, 'British merchant shipbuilding, c. 1500-1750', *International Journal of Maritime History*, 11, 1999, 27-52, at 36.

56 Pett, *Autobiography of Phineas Pett*; Roy McCaughey, 'Phineas Pett, (1570-1647), shipbuilder and naval administrator', *Oxford Dioctionary of National Biography*; National Maritime Museum, SOC/15, fos. 21r, 43r, 64r, 84r, 149r.

57 J K Laughton (rev. J D Davies), 'Peter Pett (*b*. 1610, *d*. in or before 1672), naval administrator', *Oxford Dictionary of National Biography*; London Metropolitan Archives, P93/DUN/327, fos. 106r-v.

58 Andrews, *Ships, Money & Politics*, 195. On Moulton, see below; on Swanley, M L Baumber, 'An East India captain: the early career of Captain Richard Swanley', *The Mariner's Mirror*, 53, 1967, 265-79; and *idem*, 'Richard Swanley (1594/5-1650), naval officer', *Oxford Dictionary of National Biography*.

59 BL, Add. MS 9306, fos. 101r, 109r; TNA, ADM 7/673, p. 245.

60 Gentles, *English Revolution*, 334-5.

61 *To the right honourable the lords and commons assembled in parliament, at VVestminster. The humble petition of the knights, gentry, clergy and commonalty of the county of Kent* (1648); for an account of the Kent petition by one of its proponents, see M[atthew] C[arter], *A most true and exact relation of that as honourable as unfortunate expedition of Kent, Essex and Colchester*, 1650.

62 Dorothy Gardiner, ed., *The Oxinden and Peyton letters, 1642-1670*, Sheldon Press, London, 1937, 138.

63 BL, Add. MS 9305, fos. 71r-4r.

64 Historical Manuscripts Commission, *Portland MSS*, I, 460.

65 *Ibid*, 460; BL, Add. MS 9300, fo. 83r, Add. MS 9305, fo. 71r. On these small vessels, see TNA, SP 16/518, fos. 6r, 24r.

66 BL, Add. MS 9305, fo. 8r; TNA, ADM 18/3, fo. 38r.

67 BL, Add. MS 9305, fo. 69v, 73v-4r; on Moulton, see Bernard Capp, 'Moulton, Robert (c.1591-1652), naval officer', *Oxford Dictionary of National Biography*; Hall had been accused of speaking against the king in 1646, TNA, ADM 7/673, 2, 32, 40.

68 BL, Add. MS 9305, fos. 71v-72v.

69 Historical Manuscripts Commission, *Portland MSS*, 461

70 *Ibid*, 460-1.

71 *MP*, 9, 23-30 May, 1648, 3 [sig. I2r].

72 *Journal of the House of Commons*, v, 576.

73 *Journal of the house of Lords*, x, 286.

74 Anon, *The declaration and propositions of the navie with the oath which they have taken, concerning an admirall for the seas, and who they have made choice of for the present*, [1 June] 1648, 1. On Kem, see Barbara Donagon, 'Kem, Samuel (1604-1670), Church of England clergyman and army officer', *Oxford Dictionary of National Biography*. Kem had preached to the king in 1647, encouraging a peaceful resolution to negotiations: Samuel Kem, *An olive branch found after a storme in the northern seas*, 1647.

75 Anon, *The declaration and propositions of the navie*, 2.

76 *MI*, 167, 25 May-1 June, 1648, 11 [1371] reported five ships involved in the revolt; *MP*, 9, 23-30 May, 1648, 8 [sig. I4v], mentioned 'a whole *Squadron* of ships'.

77 Anon, *The declaration and propositions of the navie*, 3.

78 *Journal of the House of Commons*, v, 606.

79 Anon, *The declaration and propositions of the navie*, 4.

80 *The declaration of the navie, being the true copie of a letter from the officers of the navie, to the commissioners; with their resolutions upon turning out Colonell Rainsbrough from being their commander. 28th May 1648*, 1648.

81 For 1648 petitions see Ashton, *Counter-revolution*, 139-58.

82 *The declaration of the navie: vvith the oath taken by all the officers and common-men of the same…May, 28th 1648*, 1648.

83 *The declaration of the navie*.

84 C[arter], *A most true and exact relation*, 52; for the prominence of Rainsborough, see Kennedy, 'The naval revolt of 1648', *English Historical Review*, 77, 1962, 247-56, at.253; Powell, *The Navy in the English Civil Wars*, 156-7; Capp, *Cromwell's Navy*, 26. For more general accounts which focus on Rainsborough, see Ashton, *The English Civil War*, 323-4; Bennett, *The Civil Wars*, 300; Woolrych, *Britain in revolution*, 411; Braddick, *God's Fury, England's Fire*, 540; Worden, *The English Civil Wars*, 97; Miller, *The English Civil Wars*, 181. This interpretation goes back to Gardiner and Ranke: Leopold von Ranke, *A History of England Principally in the Seventeenth Century*, 6 volumes, Clarendon Press, Oxford, 1875, II, 516; Gardiner, *History of the Great Civil War*, IV, 135.

85 Anon, 'The reasons [th]e Navy give for theire Resolution', dated 17 June 1648, E.448[3]; BL, Add. MS 9305, fos. 4v, 43v, 47r.

86 *PD*, 253, 29 May-6 June, 1648, 6 [2039]; *MI*, 168, 1-7 June, 1648, 3 [1375].

87 On the siege, see Braddick, *God's Fury, England's Fire*, 545-8.

88 Historical Manuscripts Commission, *Portland MSS*, 462.

89 *Journal of the House of Commons*, v, 606.

90 *Journal of the House of Lords*, x, 297-300.

91 BL, Add. MS 19,367, fo. 3v; *LJ*, x, p. 313.

92 TNA, ADM 18/3, fo. 37v; the letter is undated but must have been written while Warwick was at Portsmouth, in early June. The *Anthelop*'s departure is also reported in *Mercurius Elencticus*, 29, 7-14 June, 1748, 2-5 [225-6].

93 Batten, *A Declaration of Sir William Batten*, 3.

94 Anon, *A fight at sea between the parliament ships & those that revolted* [30 June] 1648, 2-3; Powell, *The Navy in the English Civil War*, 165, gives 10 June, but seems to have some confusion concerning dates in June and July.

95 BL, Add. MS 9300, fo. 91r; cf *CJ*, v, 599.

96 *Journal of the House of Commons*, v, 606.

97 *Journal of the House of Lords*, x, 339-40.

98 A manuscript note entitled 'Marriners & seamen of [th]e Trinitie-house theire Resolution' and dated 21 June 1648 survives in the Thomason tracts, 669.f.12[51-2]; practically the same text was printed as *The humble tender and declaration of many well-affected mariners and sea-men, commanders of ships, members of Trinity-house, to the commissioners of the navy*, [23 June] 1648. The only surviving account of this meeting is in Richard Badiley, *The sea-men undeceived*, [18 August] 1648; the minutes of Trinity House for this period do not survive: see Harris, *Trinity House*, 4-6. For Badiley, see below.

99 *The humble petition and desires of the commanders, masters, mariners, younger brothers and sea-men of the shipping belonging to the river of Thames* [1 July] 1648; *Journal of the House of Commons*, v, 615-6; *Journal of the House of Lords*, x, 351-2.

100 *The humble petition and desires*, 6.

101 *The humble declaration, tender, and petition of divers cordiall and wel-affected marriners, whose names are subscribed, to the right honourable the lords and commons assembled in parliament* [6 July] 1648; *Journal of the House of Commons*, v, 624; *Journal of the House of Lords*, x, 363.

102 *Mercurius Elencticus*, 31, 21-28 June, 1648, 4-5 [240-1]; *Mercurius Pragmaticus*, 14, 4-11 July, 1648, 5-6 [sig. P3r-v]; *Mercurius Elencticus*, 33, 5-12 July, 1648, 2-3 [254-5].

103 *Mercurius Elencticus*, 34, 12-19 July, 1648, 7-8 [267-8].

104 *Mercurius Pragmaticus*, 17, 18-25 July, 1648, 8 [sig. R4v], 18, 25 July-1 August, 1648, 12 [sig. S6v], and 19, 1-8 August, 1648, p. 9 [sig. Xr].

105 *To the right honourable the commons assembled in parliament: the humble petition and representation of divers well-affected masters and commanders of ships*, 1648; this was presented on 11 September.

106 *Mopderate Intelligencer*, 175, 20-27 July, 1648, 2 [1458], 176, 27 July-3 August, 1648, 12 [1480]; *Mercurius Pragmaticus*, 14, 4-11 July, 1648, 6, [sig. P3v], 18, 25 July-1 August, 1648, 12 [sig S6v], and 20, 8-15 August, 1648, 9 [sig. Z1r]; BL, Add. MS 9300, fo. 95r.

107 Historical Manuscripts Commission, *Portland MSS*, 482.

108 BL, Add. MS 17,677 T, fos. 171r-v; *Perfect weekly account*, 21, 2-9 August, 1648, p. 3 [sig. W2r]; *Mercurius Pragmaticus*, 21, 15-22 August, 1648, 11-12 [sig. B2r-v].

109 *Mercurius Elencticus*, 35, 19-26 July, 1648, 5 [273] and 38, 9-16 August, 1648, p. 8 [311]. In his will Jordan, intriguingly, left 'to

soe many as were Elder brothers of the Trinity House in the yeare one thousand six hundred Fourty and eight, and shall be living at the time of my decease, A Ring of Gould of the value of Twentie shillings': TNA, PROB 11/253, proved 13 February 1656.

110 *Journal of the House of Lords,*, x, 432-4.

111 *A declaration of the officers and company of sea-men aboard his majsties [sic] ships* [8 July] 1648; anon, *The oath taken by the sea-men of the revolted ships* (15 July 1648); *The declaration of the sea commanders and marriners in the royall navie and fleet, now with his highnesse Prince Charles, riding on the Downes,* 2 August 1648.

112 For these publications, see the notes below: for Badiley, see Bernard Capp, 'Badiley, Richard (c.1616-1657), naval officer', *Oxford Dictionary of National Biography*; for his involvement in the 5 July petition, *Mercurius Elencticus*, 34, 12-19 July, 1648, 7-8 [267-8].

113 TNA, ADM 7/673, 264.

114 W[illiam] B[atten], *The sea-mans diall*, 3-5.

115 *Ibid*, 4-5.

116 Richard Badiley, *The sea-men undeceived*, 12.

117 *Ibid.*, 12-13.

118 *An ordinance of the lords and commons assembled in parliament, authorizing Robert Earl of Warwick Lord High Admiral of England, to execute marshal-law* (1648); *PD*, 265, 21-28 August, 1648, p. 5 [2134]; *Mercurius Pragmaticus*, 22 [A (there were two different texts printed as this issue)], 22-29 August, 1648, 6 [Sig. Cc3v]; *Mercurius Pragmaticus*, 24, 5-12 September, 1648, 8 [sig. F4v]; *Mercurius Melancholicus*, 54, 28 August-4 September, 1648, 6 [328].

119 Scott, *Politics and War*, 179-80; Gentles, *The English Revolution*, 339-49.

120 For a list of Warwick's fleet, *Journal of the House of Lords*, x, 495.

121 Warwick's account of these events, as written to parliament, is in *Journal of the House of Lords*, x, 488-90; it was printed in *PD*, 267, 4-11 September, 1648, 3 [2147]; a royalist account was published shortly after: anon, *The copie of a letter from a commander in the fleet with his highnesse the Prince of Wales*, [20 September] 1648; shorter accounts were presented in *Mercurius Pragmaticus*, 23, 29 August-5 September, 1648, 12 [sig. Ef2v]; *Moderate Intelligencer*, 181, 31 August-7 September, 1648, 2-3 [1518-9]; *Perfect occurrences of every dayes journall in parliament* [hereafter *PO*], 88, 1-8 September, 1648, 6 [438].

122 For events at Goree, in roughly chronological order: *Moderate Intelligencer*, 183, 14-21 September, 1648, 4 [1544]; *Mercurius Militarius*, 1, 10-17 October, 1648, 4 [4]; Anon, *A fight at sea two ships taken by Prince Charles his officers*, [7 November] 1648; *Mercurius Pragmaticus*, 32 & 33, 32 October-14 November, 1648, p. 12 [sig. Yy2v]; *Modeate Iintelligencer*, 190, 2-9 November, 1648, 9 [1735]; *Mercurius Elencticus*, 51, 8-15 November, 1648, 7 [500]; *PO*, 98, 10-17 November, 1648, 3-4 [711-2]; Anon, *Two letters containing all the proceedings bettwixt the Prince, and the E. of Warwick: a fight at sea, and a fleet upon the Downs*, 1648 [the pamphlet bears no date, but the latest letter is dated 18 November]; *MP*, 34, 14-21 November 1648, 5-6 [sig. Bb3r-v]; *Mercurius Elencticus*, 52, 15-22 November, 1648, 7 [507]; *Moderate Intelligencer*, 192, 16-23 November, 1648, 2 [1742]; Anon, *A letter from the navy vvith the earle of Warwick, lord admiral, from Hellevoyt Sluice, November.* 24, 1648; BL, Add. MS 9300, fo. 106r.

123 Capp, *Cromwell's Navy*, 59-66.

124 On 'Pride's purge', see David Underdown, *Pride's Purge: Politics in the Puritan Revolution* (Clarendon Press, Oxford, 1971).

125 *Act touching the regulating of the officers of the navy and customs*, 1649.

126 *The declaration and engagement of the commanders, officers, and seamen in the shippes, under the command of the right honourable the earle of Warwicke* [28 December] 1648.

127 For recent debates concerning Charles's trial and execution, see Clive Holmes, 'The trial and execution of Charles I', *Historical Journal*, 53, 2010, 289-316; this is a response to Sean Kelsey, 'The death of Charles I', *Historical Journal*, 45, 2002, 727-54; idem, 'The ordinance for the trial of Charles I', *Historical Research*, 76, 2003, 310-3; idem, 'The trial of Charles I', *English Historical Review*, 118, 2003, 583-616.

Richard J Blakemore *is researching for a PhD at the University of Cambridge. His thesis focuses on the London and Thames community during the civil wars.*

CHARLES, JAMES AND THE RECREATION OF THE ROYAL NAVY 1660-1665*

Hilary Todd

Abstract

This paper explores the role of James, Duke of York, brother of Charles II, in the recreation of the Royal Navy in the early Restoration period. The task facing Charles, and James as his High Admiral, was to turn the bankrupt and mutinous fleet back into a Royal Navy. The credit has usually gone to Charles aided by Pepys, but James's letter books in The National Archives show him as being involved and active, seeking advice, finding out what practice had been and trying to bring order to ships, men, victuals, and discipline. The argument here is that James was Charles's essential right-hand man, ably assisted as he was by others. He was visible and his care was noted. By 1664, with war with the Dutch looming, it was a Royal Navy that put to sea under James's own command, and initially at least, it had some success.

The Restoration was one of the most remarkable changes of regime in British history and it inaugurated a fascinating period. In the Navy, it would see the development of bigger, heavier ships and the developing use of the line of battle, in three squadrons, the basic tactic that would prevail for the next century or so. In support of this was the development of dockyards and fortifications, more attention being paid to the manning of the fleet with the development of a career structure and the requirement of examinations for commissions. Yet this is often glossed over in the popular imagination which seems to move seamlessly through a fascination with Henry VIII and Elizabeth I, to the eighteenth century and empire. However the Stuarts' period of domination was not greatly shorter than that of the Tudors, and, has been pointed out by John Miller, it was under the Stuarts, not the Tudors, that England emerged as a major European and world power.[1] In navy terms, interest has tended to move from Drake and the Armada to what has been called the Long Eighteenth Century starting in 1688,[2] to Nelson and Trafalgar. That misses a colourful and crucial period in our history, and not just for the Navy. The origins of so much that followed can be traced to the surge of energy which followed the Stuart Restoration.

In looking at the transformation of the Navy of the Republic back into a Royal Navy in the period to 1665, perhaps slightly daringly at a conference with 'Pepys' in its title, the aim of this paper is to bring to the fore the role of James, Duke of York, as Lord High Admiral. When talking about the post-Restoration Navy, the two names most associated with it have been the King, Charles II, whose navy it undoubtedly was, and Samuel Pepys. Much has been written about the King's love for his navy and deep involvement with it, and likewise Pepys, and titles of books have reflected that preoccupation.[3] But the man who was Lord High Admiral for the first thirteen years of the reign was James, Duke of York, and even after he left office, his influence could not be ignored.

James was in a unique position in holding the office, as brother of the King, member of the Privy Council and for much of the reign, heir to the throne. He was, therefore, at the heart of strategic decision-making, and alongside his brother, directed the work of the Navy Board. His was not the problem of many second sons of kings – he had been given a proper job and the contention of this paper is that for him, it was no sinecure. He was a very different character from Charles, serious, a man of action, and not a drinker or a gambler. Moreover, in the years in exile,

---

Paper presented at the fourteenth annual conference of the Naval Dockyards Society held at the National Maritime Museum, Greenwich, on 17 April 2010. Theme: Pepys and Chips, Dockyards, Naval Administration and Warfare in the Seventeenth Century.

he had forged his own career, a military man who had learned his trade in the armies of France under the great French general, Turenne. That career had been brought to an abrupt halt by his brother's restoration to the throne. It was while researching a thesis on the Navy of Charles II in the approach to the Second Dutch War that the author came to notice James's name, and a question formed as to whether his conversion to Catholicism and failure as a King had served to ensure that his part in the Restoration navy should be largely eclipsed, an eclipse rendered more complete because of the fascination with two far more accessible and successful characters, Charles himself and Samuel Pepys. The wish to revisit his role was encouraged by Andrew Lambert's inclusion of him in his book, *Admirals*,[4] and bolstered by more favourable views voiced in recent works by Professor Rodger[5] and by working on James's letter books to be found in The National Archives at Kew.

The National Archives holds in ADM2 a number of letter books of orders and letters from James's time as Lord High Admiral. For the purposes of this essay, use has been made of just two of them, ADM2/1745 Lord's Letters: Duke of York's private letter-book 1660 – 1668, and ADM2/1732: Admiralty Order Book, Duke of York, June 1660-September 1662 which are being looked at in detail. These are large books, foolscap size, and at present if you wish to consult them, you will be handed the original documents, a real treasure trove for historians.

The contention here is that there is much in these letter books to provide a balance to the historical legacy of Samuel Pepys. From Bryant onwards, his name has come to dominate any study of the administration of the Navy of this period. The ultimate self-publicist, he has achieved the status of a literary personality in our time, with his famous diary now on line complete with a Twitter account[6] and of course the vast collection of documents in his library at Magdalene College Cambridge. As Nicholas Rodger says, 'Thanks to him we know more about the administration of the Royal Navy of his day than of all earlier and many succeeding periods, but of course we see it through his eyes'.[7] In writing of the administration of the time, the challenge now is to make greater use of the Navy Board and other sources in order to achieve a more balanced view.

James, by contrast, has had a much rougher ride, dismissed by the Whig politician and historian, Macaulay [who wrote *The History of England from the Accession of James the Second* (4 vols 1848-55) in praise of the glorious revolution of 1688 and is hence very anti-James] as someone who 'would have made a respectable clerk in a dockyard at Chatham'[8] and with few friends amongst his biographers.[9] Much of the interest in James has turned on his short reign and his unsuccessful attempts to regain his throne after 1688. He was not a politician, and fatally lacked Charles's political facility, elasticity of morals and intellect and his sense of timing. But while his characteristics may have been fatal to him as king, they might just have been the ones required for the running of a navy, his brother's not entirely welcome interventions notwithstanding[10] and he also left a small body of letters. Consideration of his papers, his experiences

Fig. 1. James VII and II (as Duke of York) by Sir Peter Lely, c.1665-70. Courtesy Scottish National Portrait Gallery.

as a professional soldier and his actions puts the present writer on the side of those who would look at his role in naval affairs anew,[11] on the basis that it has deliberately been diminished, very much on the basis of history being written by the victors, many with uneasy consciences.

The period covered by this paper starts with the necessity to pay off much of the Navy inherited from the Republic and ends with the start of another conflict with the Dutch in 1664/5. This time, it would be a Royal Navy that put to sea commanded in person by the Lord High Admiral, and initially at least it had some success. There is, of course, a huge 'but' to follow that last statement, but the effects of the plague, fire and raid on the Medway belong to another paper.

The Restoration, as a number of writers have made plain, was by no means a foregone conclusion and it came about with a speed which took even its major players by surprise. In January 1660, James was being offered the position of Lord High Admiral of Spain, and Charles seemed as far away from his throne as ever.[12] Yet, through John Lawson bringing the fleet into the Thames, George Monck, his troops over the Coldstream, and correspondence from Edward Montagu and others with the royalists in exile, by April 1660, Charles had published the Declaration of Breda, Grenville's overtures had been well received by Parliament such that it had declared Charles King on 8 May and Montagu had been dispatched to transport him back to England.

It is worth reminding ourselves just how young the returning princes were – Charles was 30 the day he returned to London on 29 May 1660; James, Duke of York, was 26 and Henry, Duke of Gloucester, just 19. In other words, they were not dissimilar in age to Princes William and Harry in our own time. They were young, attractive and unmarried, and facing problems of a monumental nature. Many of those who surrounded them were much older than themselves, with experience of commanding armies, ships and fleets on both sides of the Civil War conflict. Some of them had lost everything in supporting the royalist cause, others had benefited to a similar degree and were concerned about losing it, not to mention politicians all jostling for position and favour. It was never going to be easy.

In the Declaration of Breda, Charles had promised to pay off the Army but there was in fact no mention of the Navy.[13] However, he clearly needed to establish the new order, loyal to himself, and quickly, particularly bearing in mind the discontent at lack of pay which was rampant. So how was it done?

The navy Charles inherited was very much the creation of the Commonwealth and Protectorate as Bernard Capp has described,[14] and had been used to force recognition of the new regime from those same European powers which had been expected to seek revenge following the execution of Charles I in 1649, but in fact never had. It had built a formidable reputation but following a treaty with France in 1655, money thereafter had been poured into the Army and the Navy had found itself under-funded. That situation was becoming critical by April 1659 when the Commissioners had resigned, angry that naval revenues had been diverted to the Army.[15] But there was considerable mistrust of the new order. Pepys described in his diary the reception of Charles's Declaration of Breda and the decision of the Council of War on Montagu's flagship to accept it on 3 May 1660. Although the vote was passed, Pepys says: 'Not one man seemed to say no to it, though I am confident many in their hearts were against it'.[16]

But that this was to be a very Royal Navy and a very personal concern was shown by the two major changes made by Charles even before arriving in England. Firstly, just as Cromwell had used the naming of ships to make a point, so did Charles – the *Naseby* was instantly renamed the *Royal Charles,* the *Richard* became the *Royal James,* and the *Dunbar,* the *Henry.* In all, some 27 ships were renamed and the naming of ships would continue to be the personal concern of Charles.[17] In addition, Montagu had already sent for carpenters and painters to take down the arms of the Protectorate and replace them with the Royal arms.[18] Charles and James clearly saw the ships as 'theirs'. It shows in the names of the warships and the yachts with *Royal Charles, Royal James, Katharine* and *Anne* as examples, sometimes, as in the case of the first two, offering up unfortunate hostages to fortune, the first being taken and the second burnt to the waterline in the Medway raid in 1667. Secondly, Charles confirmed James as Lord High Admiral, a position first conferred on him by his father, Charles I, in 1638 when he was a boy of 4 ½,[19] and James immediately took charge. Pepys noted in his diary that: 'My Lord doth nothing now; but offers all things to the pleasure of the Duke as Lord High Admirall - so that I am at a loss what to do'.[20]

James had a nice sense of his own position and no one could have been in any doubt about it. Many of his early letters contain the phrase, 'my Sovereign Lord and Brother', making clear he saw himself as a servant of Charles's but also very close to him. He was a man not to be trifled with. He also expected action. The word 'speedy' is a favourite as well as the phrase 'with all convenient speed' and in sailing orders, 'at the first opportunity of wind and weather'. A good example is the letter of 4 March 1660/1:

I desire you will forthwith give direction for the speedy refitting of his Ma:^{ts} shipps the Swiftsure and Colchester as part of this Sum:^{er} Guard and that all dispatch bee used therein to the end they may bee suddainly in a readinesse to put to sea.[21]

There is a sense of urgency in the orders that he gave from the start.

In terms of regime change some, of what was done was exactly what might be expected, with firstly, purges of those inimical to the new regime just as had been done in the past, largely achieved by the administration of the Oaths of Supremacy and Allegiance in the fleet and the dockyards. This was not a realm united in welcoming back its king. In September 1660, James's secretary, William Coventry, wrote to the Principal Officers concerning the officers on the *Happy Return*, clearly not a happy ship despite her name, who had refused the Oaths and so needed to be discharged. It must have thrown the ship into chaos as he then wrote to Lawson about her telling him to replace her in the convoy being readied.[22] And again, on 19 December 1660, James wrote requesting the discharge of a number of men employed in yards and ships at Chatham and Deptford who 'have refused or declined to take ye oaths of supremacy & Allegiance due to the King my Sovereign Lord & Brother'.[23] The list of 26 people included five shipwrights, a mast-maker, rope-maker, several labourers, a carpenter, two men of the *Sovereign*, most of them skilled men his Majesty's service could ill afford to lose.

There was clearly trouble or the potential for it again in January 1661. A few days after the rising of the Fifth Monarchy Men under Venner, in the putting down of which both Monck and James were involved,[24] James wrote to Sir William Compton, Master of the Ordnance: 'The late disorder here giveth me cause to judge it necessary to order some extraordinary Guard for the several yards at Deptford, Woolwich and Chatham' and he requested arms for those sent to sort it out. Sir William Batten was to go to Chatham 'to put that Yard in order' and the Comptroller of the Navy, Robert Slingsby, was to take care of the other two.[25] Portsmouth, being a garrison town, was not considered to need further reinforcement. In the event, the rising was rapidly put down, but one outcome was the realisation that the King needed at least a bodyguard, a point Haswell says James had been reluctant to press for fear of a political storm.[26]

Although some modern writers have played down the threat of unrest and insurrection, considering the early life experiences of both James and Charles, in the author's view it is not surprising that they never felt safe in the high offices in which they had been placed. Security would always be a consideration for them. Charles had no wish to go on his travels again and Haswell contended that having seen his father's hand forced by the mob to sign Strafford's death warrant for want of a bodyguard, James would always want troops for protection.[27] But many would accept the change of regime and just get on with life in the hope that calm would be restored. The Pett family of naval administrators at Chatham has been quoted as one such 'accepting Charles II as readily as it had abandoned his father'.[28]

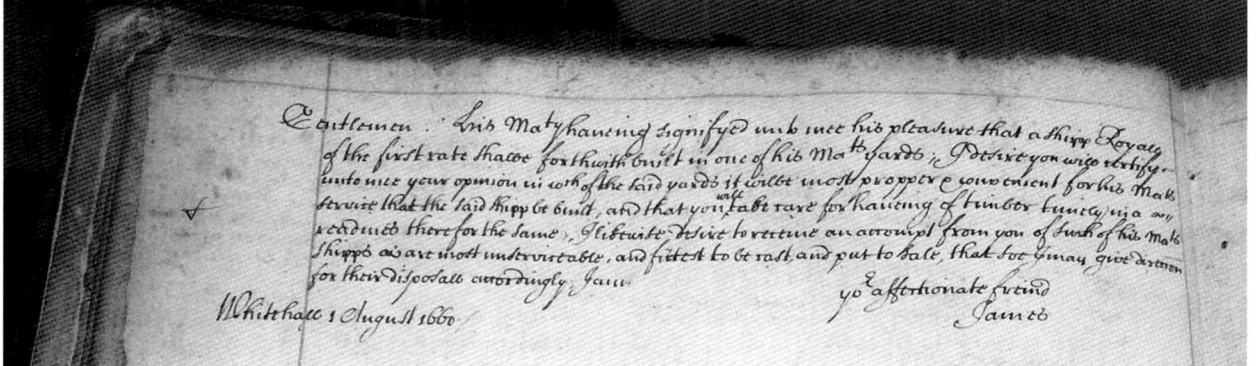

Fig. 2. James's letter to Principal Officers of the Navy Board requesting the construction of a new first rate ship, 1 August 1660. Courtesy of The National Archives.

The paying off of the Navy was a key task alongside that of continuing the normal business. So as might be expected, one of the first concerns for James as Lord High Admiral was money. Almost as soon as he had returned to London, he had dispatched George Carteret, not yet Treasurer of the Navy, to visit the office ('I desire you to repair to the office where the affaires of the Treasury are managed and examine the present state of the Treasure').[29] He was concerned that the 'present posture of the affaires of the Navy is very intricate, and the debts great' and wanted to know exactly what had been allocated 'for the support provision and pay of the navy' and to work with the existing Treasurer and report to him. He recognised that 'without speedy remedy' his Majesty's service was likely to be very much 'endamaged'. By August, he was writing to the Clerk of the Cheque at the various dockyards about seamen who could not be paid because of 'the present want of moneys' and directed them not to discharge mariners or seamen except for sickness or other pressing necessity.[30] Money or the lack of it would be a constant theme in these early years, and runs like a thread through the letter books. It took far longer and far more of it to pay off the debts than anyone could have thought. Nevertheless, the fleet was finally paid off in what Knighton describes as an 'oasis of solvency' in 1663/4.[31] But money would limit what could be done in this early period.

Within a couple of months of arriving, Charles, always conscious of the power of show, evidently wanted to emulate his father in building a new first rate ship. Then as now, there was nothing like a big, highly decorated warship to attract attention and project the power of the regime that had built it, both at home and abroad. Cromwell had known that, and built the *Naseby*, and Charles no doubt wanted to show that he, too, could do the same. As early as August 1660, James wrote, perhaps a little pompously but with obvious pride, to the Principal Officers

> His Ma^{ty} haveing signifyed unto mee his pleasure that a Shipp Royall of the first rate shalbe forthwith built in one of his Maties yards

and asked for advice on which yard it could be built in (see Fig. 2).[32] But the exercise would also demonstrate the disabling lack of money as after a couple more letters about timber and where best to build the ship, the subject disappears and other records show a new first rate would not be built until later in the decade. However, the issue of the first rates did not altogether disappear as on 8 August, James gave orders for the *Sovereign*, their father's first rate, launched in 1637, to be floated out at Chatham and the *Prince* brought in for repairs.[33] Instead, in 1661, two second rates were ordered,[34] but even these were not completed until 1664. These ships fulfilled the role of prestige, projection of power of the regime and reflected the developing tactics of the time in terms of fleet actions.

However, if first rates were beyond the purse of the new regime, very small ships were not. Early on, James ordered the laying up of one little vessel especially dear to Charles's heart, the smack, the *Surprise*, in which he had made his escape after the battle of Worcester in 1651, and which he had purchased. It was thereafter renamed the *Royal Escape*, her boatswain paid as of a fifth rate despite her very small size, and she was opened to distinguished visitors.[35] In addition, the brothers took great delight in their yachts, small ships Charles had first encountered in Holland, described by John Evelyn as 'very excellent sailing Vessels'.[36] In August 1660 he had been given his first, the *Mary*. Of this yacht, James requested of the Principal Officers

> *I desire you will forthwith consider of some convenient place neere Deptford where his Ma^{ts} yacht may continually lye afloate, to the end she may be in a readines upon all occasions for his Ma^{ts} Service, and to take care that as she lyes she may be secured from receiving any prejudice by the boates & vessels passing upp & down the River.*[37]

Not to be outdone, James also had a yacht built, *Anne*, named after his wife.

A number of descriptions of races between the brothers are to be found in both Pepys and John Evelyn's diaries.[38] This was sibling rivalry played out on water and a very personal use of navy ships. However, although designed for pleasure, cruising and racing, they were full members of the fleet, highly decorated small warships armed with four to eight guns and sumptuously fitted out and with vast silk flags. Fox described them as 'excellent wartime scouts and advice boats'.[39] Both Charles and James had given standing instructions that theirs should be available at a moment's notice and they were used by both on their visits to the fleet. They were also used for conveying people in style, but also by other members of the administration. Pepys describes one such voyage on the *Bezan* on 17 August 1665, even going to sleep 'upon ve[l]vet cushions of the King's that belong to the Yacht'.[40] They could also be with the fleet in battle, the *Henrietta*, built in 1663, was sunk in action in 1673 and earlier, in March 1666, James requested a 'free Guift' of money for Captain Fasely of his yacht, *Anne*, 'in consideration of his losse of his Legge in ye Kings service'.[41]

Another example of the quasi-state use of ships was in the protection of trade and the gaining of more of it. Thus besides the orders for fishery and trade protection, James had no hesitation in ordering the fitting out of ships to accompany expeditions to Guinea, and to America to challenge the Dutch monopoly. In an age of mercantilist thinking, in which trade was thought to be finite, the only way to get more was to take it off someone else, in this case the Dutch, and in any case James had an interest in the Royal African Company. More modern eyes might view this as a conflict of interest but it did not seem so to James and others of his circle who shared his interest. A much larger expedition of twelve ships was put in hand in August 1664.

In terms of the structure of the administration, it was the recommendations of a committee he himself chaired that James put forward to the Privy Council and which were duly accepted on 4 July 1660. It could be argued that none of the ideas were those of James himself, but that he merely presented them. But that does not chime with what is known of his history and character. He had to set out the recommendations, and took responsibility for their implementation. Perhaps unsurprisingly, given the nature of the times and the need for speed, the structure proposed was a return to the Navy Board with the ancient offices associated with it. It is generally accepted that the men chosen to fill the positions, were, with a couple of exceptions, men of experience in naval matters and an example of the balance Charles (and James) were trying to strike in the appointments made. Thus the Board comprised: the four Principal Officers (Sir George Carteret – Treasurer of the Navy, a former Vice-Admiral for Charles I; Sir William Batten – Surveyor of the Navy; Sir Robert Slingsby – Comptroller of the Navy (replaced on his death in 1661 by another sea officer, Sir John Mennes), and a young Samuel Pepys as Clerk of the Acts whose place was obtained through the patronage of Edward Montagu.

In a departure from previous practice, to them were added three Commissioners without portfolio as had been done during the Interregnum: Peter Pett, Sir William Penn, a former General-at-Sea, and Lord Berkeley of Stratton, another veteran royalist and the other besides Pepys without sea experience.[42]

Pepys's diary indicates that the members of the Board met with James as Lord High Admiral to do business[43] (it also shows that in this period, Pepys, who was just a few months older than James, was finding his feet, learning his trade and trying his best to up his position in the eyes of the rest of the Board!). But as the letter books show, the orders start with James (and sometimes William Coventry, his secretary and from 1662, a member of the Navy Board) and the Principal Officers were expected to carry them out. All this activity was played out against a backdrop of tremendous faction and in-fighting.

In terms of the key men, those who had assisted in the King's return were amply rewarded. George Monck became Duke of Albemarle, Edward Montagu, Earl of Sandwich. But others of the old regime were also knighted in the early post-Restoration years, men such as William Penn, John Harman, Christopher Myngs and John Lawson.[44] However great the debt owed by Charles to royalists who had sacrificed everything in his service, he could not afford to alienate the men with experience. James also knew that – his letters to Montagu and to Lawson show him making overtures and a willingness to seek their advice.

The early correspondence shows that James was very much feeling his way and was alive to the concerns of those who were working with him. He was also keen to establish good relations with them. But as time goes by, the tone shows increasing confidence as he mastered the terminology and the names of ships and people. He was clearly putting time and thought into what he was doing. Contrary to the impression given that James never took advice, at this stage of his career, he was actively seeking it. There is a much quoted passage in Pepys's diary which refers to him having expressed his willingness to 'learn the seaman's trade of him [Montagu], in such familiar words as if Jack Cole and I had writ them'. And it is clearly with Montagu that he has been in correspondence as Montagu admits to Pepys in the same passage.[45]

In a letter to Montagu on 6 June,[46] just a few days after Charles's arrival in London, James said he had passed his letter to the Committee of the Admiralty, then still in being, for their views and he relayed what they advised. But he also referred to an earlier letter he had sent to Montagu from which he would know how 'desirous I am to discourse with you' about the state of the Navy.

Admiral Lawson had to wait a little longer for his response. James wrote to him on 14 June in answer to his two letters but explained that he had been waiting for advice from the Committee of the Admiralty about what had been the usual practice in relation to the handling of prizes.[47] But thereafter, his correspondence with Lawson developed. As he did not yet know what ships were in the Downs, he thought 'fit to referre it unto you to direct' a fourth rate to accompany a ship to Dordrecht in Holland ('they alleadging it customary to have one of that quality'). There are several where he has been asked to find transport for people on official business or just to transport them to Holland.[48]

Equally understandably, Charles and

Fig. 3. 'The Grand Seale for the Lord High Admiral JAMES DUKE of YORK, Brother of KING CHARLES 2d'. Courtesy National Portrait Gallery.

James wanted to promote the interests of 'Cavaliers' such as Sir Robert Holmes and Sir Edward Spragge. The issue of 'gentlemen and tarpaulins' and the factions associated with them so comprehensively described by David Davies would be an ongoing issue,[49] but the key concern for the royal brothers was to hold all the interests in balance as best they could in the highly volatile circumstances of the time.

Some of the problems in terms of personnel were of the royal brothers' own making. Despite the recall on 7 June 1660 of privateer commissions issued by both Charles and James whilst abroad,[50] James had to write twice to one, the notorious Richard Beach:

I am informed that not withstanding his Majs Restoration to his Kingdomes and Proclamation made thereof on the 8th day of May last, you continue to act in an hostile manner against his Maj's subjects under Pretence of a Commission derived from me…[51]

Beach clearly did obey orders in the end and indeed was one of the few cavaliers to secure a captain's commission in the early period.[52]

In terms of the future, however, in early 1661, there is a letter from James concerning the first of what would be known as the King's letter boys, a scheme for encouraging the younger sons of gentry to be sent to sea to learn to be future officers. But James was no desk-jockey. He was a visible commander, and like his brother paid many visits to the fleet. And his care was noted. The problem for a navy, then as now, was that the service had to expand and contract as its political masters required, in this case the King but with the support of Parliament which had to vote the money.

By 1664, relations with the Dutch were on the slide, and as early as May 1664, James was ordering ships to sea manned 'as in time of Warre' in anticipation. He was also, in August, asking members of the Board to go to Harwich to look at how best the house and yard there, which had been used for cleaning and refitting ships, might be brought back into use which he clearly thought would be needed as indeed it was. As might be expected, they were to report their findings to him with 'all convenient speed'.[53]

In November 1664, as is well known, Parliament voted £2.5 million for the impending war, the largest supply ever voted to an English monarch to that time.[54] Its passage was described in a letter from Thomas Clifford to William Coventry dated 25 November 1664, which included a tribute to the Duke of York: 'I assure you his beeing at the fleet and the great care he takes of it hath bin noe small advantage to the attaining this great supply.' However it was followed immediately by a memorandum which explained why such a sum would not be sufficient. The cost of the men at sea alone would soak up most of it, and it then went on to list the headings of other additional expenditure including salaries of officers on shore and buildings, rewards for extraordinary services, recompense for widows and orphans, sick and wounded, maintenance of prisoners and, of course, the building of more ships.[55] Parliament always seemed (or indeed seems in our own time) to be astounded by the cost of maintaining a navy, largely through a failure to understand the scale of the task of supporting one.

It is clear from some of James's letters that he was well aware of the need to care for his seamen, as indeed his colleagues were also. Victualling could (and can) make or break a fleet and James knew this. The diet produced the very considerable number of calories needed for men to be able to do their heavy physical work, even if it was pretty unpalatable to modern tastes. Charles and James even seem to have tried it, as Pepys noted in his diary before the voyage back to England:

there being set some Shipps diet before them, only to show them the manner of the Shipps diet, they eat of nothing else but pease and pork and boiled beef.[56]

In the setting out of the Winter guard in November 1660, James asked the Principal officers 'with all convenient speed' to tell him what proportions of victuals were necessary for the victualling of that and the next Summer's guard and to give timely orders, distinguishing the two.[57] He wanted to know how it all worked. The letter book is full of orders for victuals for ships and for making them up where they were short. On 12 November, he had heard from Lawson that the *London*, *Swiftsure* and *Breda* had supplies for only a matter of weeks and as they were to be part of the Winter guard, he wanted them to have a further portion of two months' victuals at the least.

On 11 November 1664, he wrote at length to Mr Secretary Bennett following a visit to the fleet at Portsmouth in which he says his Majesty's affairs he finds 'to be in pretty good forwardness excepting seamen and somewhat of the victualling', not enough to prevent a good number of ships from sailing. He was more concerned about the lack of manpower, without more of which he feared

that either some of the shipps must be left in harbour ...or else they wilbe soe slenderly manned as that they will not be able to doe the service which might reasonably be expected from such shipps.[58]

He had been to see for himself, and he was well aware that without sufficient men, and those properly provisioned, the ships could not perform.

There is also evidence of James doing what he could to combat drunkenness in the fleet. Being himself of moderate habits where drink was concerned he clearly did not like excess in others, especially when it led to dereliction of duty. In 1664, when told of a 'great neglect of duty among the officers & others' of the ships and yards 'through drunkenness also other debauchery' he directed the Principal Officers to take action to restrain it so that 'no prejudice may thereby happen to his Maties service ...'.[59] Similarly, when told that the boatswain of the *Sorlings* had been so negligent in his duty and so frequently absent from the ship when in Yarmouth that it had in fact sailed without him, he had no hesitation in dismissing him and replacing him with someone of whom he had heard good testimony.[60]

The liking for order was reflected in his approach to battle tactics, just as it was in administration. It may be argued that in acting as fleet commander, he took a lead from what he had learned when serving with the great French General, Turenne, and was keen on having his commanders meet him on board the flagship to discuss tactics.[61] He was interested in how to operate a fleet at sea, issuing Sailing and Fighting Instructions for the better ordering of the fleet based on but developing those issued in the first Dutch War.[62] Again, it may be argued that they drew on the ideas of other, more experienced seamen, but equally it may be argued that as a man who was keen to meet and discuss with his commanders, his would be the last word. They were issued in his name and he took responsibility for them. And it is also clear that there was no unanimity at this point in time on whether line of battle or a simple charge at the enemy was the better approach to a fleet action. He himself favoured order, while others such as his dashing cousin, Prince Rupert, did not.[63] History would suggest that what James did in building on the original Sailing and Fighting Instructions was another step on the main line of development of fleet tactics.

As an active commander, he was considered to have acquitted himself well at Lowestoft in the opening battle of the Second Dutch War. He had at his side on the flagship the experienced William Penn as captain, and was, as commander-in-chief, a target for the enemy. But he stood his ground when his three companions, Charles Berkeley Earl of Falmouth, Lord Muskerry and Roger Boyle, were cut down by chain shot, leaving him spattered with their blood and brains.[64] He would command again at Sole Bay in the Third Dutch War and there be described by Sir John Narborough as a good man to have beside you in a fight:

The Duke thought himself never near enough to the enemy ... to say all, he is everything that man can be, and most pleasant when the great shot are thundering about his ears.[65]

It might have been said with an eye to currying favour with James but it comes in Narborough's general description of the fight and with an energy that seems real.[66] After both of these actions, James was ordered ashore by both King and Parliament, anxious that the heir to the throne should not put himself at further personal risk. His frustration at such a turn of events is evident in the stream of instructions in the letter books to those who took over command.

Finally, it is clear that James took great pride in the achievements of the fleet and its officers. It is him we have to thank for the magnificent *Flagmen of Lowestoft* series of paintings. Although it is debated as to how much of a victory Lowestoft actually was,[67] it was clear James considered it so. The portraits were commissioned to celebrate his flag officers' success at Lowestoft from one of the foremost portrait painters of the day, Sir Peter Lely, and they may be most easily accessed on line, that of Sir John Harman, the captain of James's own flagship being considered 'perhaps the most dashing' according to the commentary.[68]

So in conclusion the contention is that, in the Restoration, which came about so suddenly, two young men came to the fore who would act to turn the Navy back into a Royal Navy, by making use of what had worked in the past in case of the structure of the administration, and of the men of experience to command it. The lack of money in this period limited what they could do in the building of the big warships they wanted, but their intense personal interest meant it had attention paid to it which in the short term meant it was fit to fight. What resulted was a very Royal Navy, one for business in all senses of the word, but also pleasure.

But James as Lord High Admiral deserves more attention. He was the link between the strategy provided by the King in Council and the administration, uniquely placed as brother and heir to exert influence. And his care of the Navy, its ships, personnel and administration, was personal and noted. The fleet largely created by the Republic sailed in 1665 as the Royal Navy under the personal command of James, Duke of York and Albany, Lord High Admiral of England.

References

1. J Miller, *The Stuarts*, Hambledon Continuum, 2004 (hardback), 2006 (paperback), p xi.
2. P Le Fevre and R Harding, *Precursors of Nelson: British Admirals of the Eighteenth Century*, Chatham Publishing, 2000, preface xi.
3. Frank Fox, *Great Ships: The Battlefleet of King Charles II*, Conway Maritime Press, Greenwich, 1980. J D Davies, "A Lover of the Sea and Skillful in Shipping': King Charles II and His Navy', *Royal Stuart Society Papers* XLII, 1992, 1-16. Arthur Bryant, *Samuel Pepys: the Man in the Making: The Years of Peril; The Saviour of the Navy*, 3 vols, London, 1933-38. J D Davies, *Pepys's Navy: Ships, Men & Warfare 1649-1689*, Seaforth Publishing, 2008.
4. A Lambert, *Admirals: The Naval Commanders Who Made Britain Great*, Faber and Faber, 2008.
5. N A M Rodger, *The Command of the Ocean: A Naval History of Britain volume Two, 1650-1815*, London 2004, 96.
6. http://www.pepysdiary.com/ and http://twitter.com/samuelpepys accessed 5.12.2011
7. Rodger, *Command of the Ocean*, 97.
8. J R Tanner, 'The Administration of the Navy from the Restoration to the Revolution', *The English Historical Review*, 12, No 45,18; also http://www.knowledgerush.com/paginated_txt/1hoej10/1hoej10_s1_p582_pages.html p 582/952 *The History of England from the Accession of James II*, Vol 1 by Thomas Babington Macaulay.
9. See most recently, J Callow, *The Making of King James II: The Formative Years of a Fallen King*, Sutton Publishing, 2000, and *King in Exile*, Sutton Publishing, 2004.
10. Davies, *Pepys's Navy*, 26.
11. a view confirmed by Peter le Fevre's review of J Callow, *The Making of James II: The Formative Years of a Fallen King*, in *The Mariner's Mirror [MM]*, 87(4), November 2001, 494-5.
12. J Miller, *James II*, First published 1978. this edition: Yale University Press, New Haven & London, 2000, 24; for surprise, see also J Callow, *The Making of King James II: The Formative Years of a Fallen King*, Stroud, 2000, 86.
13. http://www.historylearningsite.co.uk/text_declaration_breda.htm accessed 17.3.2010.
14. B Capp, *Cromwell's Navy: the Fleet and the English Revolution 1648-1660*, Oxford University Press, 1989; paperback 1992.
15. *Ibid.*, 341.
16. *The Shorter Pepys, selected and edited by Robert Latham from The Diary of Samuel Pepys, a new and complete transcription edited by Robert Latham and William Matthews*, Penguin Books, reprinted in Penguin Classics, 1993, 3 May 1660, 39. [*The Shorter Pepys*].
17. Frank Fox, *Great Ships*, 57.
18. *The Shorter Pepys*, 11 May 1660, 43.
19. *Calendar of State Papers Domestic*, Charles I, Vol 12, 1637-8, 7 April 1638, p 351; For its renewal, *The Shorter Pepys*, 16 May 1660, 45.
20. *The Shorter Pepys*, 21 May 1660, 48
21. ADM 2/1745 James to Principal Officers, 4 March 1660/1.
22. ADM 2/1745, Coventry to Principal Officers, 20 September 1660, and Coventry to Lawson 26 September 1660.
23. ADM 2/1745 James to Principal Officers, 19 December 1660.
24. J Haswell, *James II: Soldier and Sailor*, Hamish Hamilton, 1972, 140-1.
25. ADM 2/1745 James to Sir William Compton, 11 January 1660/1 and *Everybody's Pepys*, 7 January, p 64 onwards.
26. J Haswell, *James II: Soldier and Sailor*, 141.
27. *Ibid.*, 4.
28. B Capp, *Cromwell's Navy*, 372.
29. ADM 2/1745 James to Sir George Carteret, 11 June 1660.
30. ADM 2/1745 James to Clerk of the Cheque at Deptford, 9 August 1660. The same to go to Clerk of the Cheque at Woolwich, Portsmouth and Chatham.
31. C S Knighton, *Pepys and the Navy*, Sutton Publishing, 2003, 43.
32. ADM 2/1745 James to Principal Officers, 1 August 1660.
33. ADM2/1745 James to Principal Officers, 8 August 1660 The *Royal Prince* had been built for James I, and the *Royal Sovereign*, for Charles I: see F. Fox, *Great Ships*, 31-7.
34. PC 6/1, 2 October 1661, 23.
35. ADM 2/1745, James to Principal Officers, 23 November 1660, 16 and M Drummond, *Salt Water Palaces*, 1979, Debrett's Peerage, 11.
36. E S de Beer, *Diary of John Evelyn*, Vol 3, 1 October 1661, 296.
37. ADM 2/1745 James to Principal Officers, 9 March 1660/1.
38. *Diary of John Evelyn*, Vol 3, 1 October 1661, 297 as an example.
39. Frank Fox, *Great Ships*, 22 comment on the *Kitchen* and *Charles* yachts.
40. *The Shorter Pepy's*, 17 August 1665, 516; see also Drummond, *Salt Water Palaces*.

41. Fox, *Great Ships*, list of yachts in service and their fates 1660-1685, 179; ADM 2/1745 James to Principal Officers 13 March 1666.
42. David Davies, *Pepys's Navy*, 26. Rodger, *Command of the Ocean*, 96-7.
43. *The Shorter Pepys*, 14 August 1660, 148-9, *Everybody's Pepys*, 3 October 1660, 53 for examples.
44. R Hainsworth, and C Churches, *The Anglo-Dutch Naval Wars 1652-1674*, Sutton Publishing, 1998, 113.
45. *The Shorter Pepys*, 3 May 1660, 40.
46. ADM 2/1745, James to General Montagu, 6 June 1660, Whitehall.
47. ADM 2/1745, James to Sir Admiral Lawson, 14 June 1660.
48. ADM 2/1745, James to Sir Admiral Lawson, 14 July 1660, and for example 13 July 1660 and 4 August 1660.
49. J D Davies, *Gentlemen and Tarpaulins: The Officers and Men of the Restoration Navy*, Clarendon Press, Oxford, 1991.
50. PC 6/1 Orders in Council, 7 June 1660.
51. ADM 2/1732 letters of 9 and 21 June 1660.
52. C S Knighton, *Beach, Sir Richard (d. 1692)* http://0-www.oxforddnb.com.lib.exeter.ac.uk/view/printable/66460
53. ADM 2/1745 James to Principal Officers, St James's 12 August 1664.
54. Hainsworth and Churches, *The Anglo-Dutch Naval Wars*, 108.
55. ADD 32094 Malet Collection, State Papers and Historical Documents, 1087-1762 Vol IV Microfilm M2670, Papers of William Coventry.
56. *The Shorter Pepys*, 25 May 1660, 50.
57. ADM 2/1745, James to Principal Officers, 5 November 1660.
58. ADM 2/1745 James to Mr Secretary Bennett, 11 November 1664, Portsmouth.
59. ADM 2/1745 James to Principal Officers, 20 April 1664.
60. ADM 2/1733 James to Principal Officers and Commissioners, 7 October 1662.
61. Haswell, *James II*, 73. Rodger, *Command of the Ocean*, 68-9.
62. B Tunstall, ed., N Tracy, *Naval Warfare in the Age of Sail: the Evolution of Fighting Tactics 1650-1815*, Conway Maritime Press, 1990, 22-3.
63. Lambert, *Admirals*, 94.
64. Hainsworth and Churches, *Anglo-Dutch Naval Wars*, 125.
65. Tunstall, *Naval Warfare in the Age of Sail*, 34.
66. See also the views expressed in Florence Dyer, 'Captain John Narborough and the Battle of Solebay', *The Mariner's Mirror*, XV, 1929, 222-32, 231 in support of this idea.
67. Tactically it was considered a victory but strategically, the Dutch fleet remained in being.
68. http://www.nmm.ac.uk/collections/explore/object.cfm?ID=BHC2750 accessed 6.11.2011.

Hilary Todd *is engaged on post-graduate studies at the University of Exeter.*

THE *LONDON* OF 1656: HER HISTORY AND ARMAMENT*

Frank L Fox

Abstract

This paper traces the previously little known history of the second-rate London, which blew up by accident in the Thames in 1665 with terrible loss of life. It explores conditions aboard the ship that might have left her vulnerable to disaster, and recounts the salvage efforts undertaken over the centuries, including a controversial modern episode. A reconstruction of the London's frequently changing armament during her short career may assist archaeologists in their searches of the river bed, and also demonstrates the not always obvious factors and policies (some unique to this one brief period) involved in arming large English warships during Protectorate and early Restoration times.

Introduction

On the 7th of March 1665 the second-rate *London* weighed from her mooring off Chatham and, guided by a river pilot, proceeded down the Medway on the ebb. The *London*, one of the most important warships in the Royal Navy and always favoured as a flagship, intended to sail a short distance up the Thames to the reach known as the Hope, where she would be joined by her new captain, Vice-Admiral Sir John Lawson. Lawson, a highly regarded sea-officer with much successful service under the Commonwealth and Protectorate, had been designated by the Duke of York, the Lord High Admiral, as the commander of the van division of his own Red Squadron in the great fleet then preparing for battle to carry on the just-declared Second Anglo-Dutch War. The ship was well short of her 450-man complement, but nevertheless had aboard at least 325 seamen. Many of these were Lawson's loyal followers and relatives who had gone with him from command to command during his career. Also present was an indeterminate number of the men's wives and sweethearts, which was a perfectly normal condition for ships in port. In vessels in which the crew had recently been paid, the term 'wives' was often a euphemism for prostitutes, but this is unlikely to have been the case in the *London*. Though not yet ready for battle, the ship was already fully armed, and her ammunition included about 300 barrels, or over 13 tons, of corned gunpowder.

Fig. 1. The *London* in 1660, by Willem Van de Velde the Elder. Courtesy National Maritime Museum, London.

On emerging into the Thames, the *London* perhaps waited briefly for the flood tide to begin and then started upstream. Having passed west of the Buoy of the Nore, she reached a position near midstream north of the mouth of the Medway when, without warning, the 13 tons of gunpowder blew up. The blast obliterated the bow section where the powder room was located, hurling massive cannons into the sky. The rest of the hull, from the waist aft, sank in seconds in about 38 feet of water so that only the quarterdeck and poop remained exposed.

Loss of life was horrific. Samuel Pepys, Clerk of the Acts of the Navy Board, entered in his diary that 'about 24 and a woman that was in the roundhouse and coach saved; the rest, being above 300, drowned'.[1] Two days

---

*This paper was not presented at the fourteenth annual conference of the Naval Dockyards Society held at the National Maritime Museum, Greenwich, on 17 April 2010, but its clear relevance justifies its inclusion in Transactions.

later, Dutch ambassador Michiel van Goch reported to his government that only 19 survived out of 351,[2] and still later the Earl of Sandwich gave the number of survivors as 'not above 12'.[3] These differences may have merely reflected mortality after the event, for the people who lived through the explosion were seared by the fireball. The diarist John Evelyn, one of the Commissioners for Sick and Wounded, went down on the 9th 'to receive the poor burnt Creatures that were saved out of the *London* fregat'. They were probably deafened as well as burned. When the Dutch flagship *Eendracht*, similar to the *London* in armament, blew up during the Battle of Lowestoft in June 1665, the noise broke windows in Dunkirk and was audible in The Hague some forty-five miles away. The death toll in the *London* was even worse than reported because of the unknown number of women aboard. Evelyn also recorded that aside from those killed, the accident left '50 widdows, & of them 45 with child'.[4]

Gunpowder explosions tended to destroy the evidence of what caused them. A correspondent of Secretary of State Henry Bennet, however, had no doubt that 'The blowing up of the *London* was caused by Chapmen selling powder 20s. a barrel cheaper than in London'.[5] But powder quality was carefully controlled. A more likely cause was an accident or negligence involving the powder room lighting arrangements, possibly during the filling and stowing of the paper or canvas cartridges used to charge the guns; cartridges had to be filled on board because they had only a few weeks of 'shelf life' before the constituents of the powder began to separate.

In English warships of the Restoration period, the main (or lower) powder room, for bulk stowage in barrels, was deep in the hold immediately abaft the foremast. Over this compartment was the 'forward platform', a part of what from about 1670 (though not before) was called the 'orlop'. Aside from storerooms for the boatswain and carpenter, the forward platform had a gunner's storeroom, which was sometimes called the 'upper powder room' because it contained low chests for stowage and issue of filled cartridges. Against one side a small 'filling room' was sunken three feet or so into the main powder room and presumably opened into it through a low bulkhead scuttle.[6] Since the slightest spark could ignite the powder, all of these compartments required some manner of protected lighting, behind heavy 'Muscovia glass'. A light room (at first called a 'fire room') was introduced only around 1690. Before that time there are only a few tantalizing glimpses of lighting customs: in March 1664[5] a list of gunner's stores requested for the first-rate *Royal Prince* included 'for ye powder roomes two large Globe Lanthorns' and 'Wax Candles for ye powder Roomes likewise'.[7] How these were placed is anyone's guess; perhaps as caged top lights extending downwards from the platform, and glazed candle niches in the storeroom bulkhead accessible only from

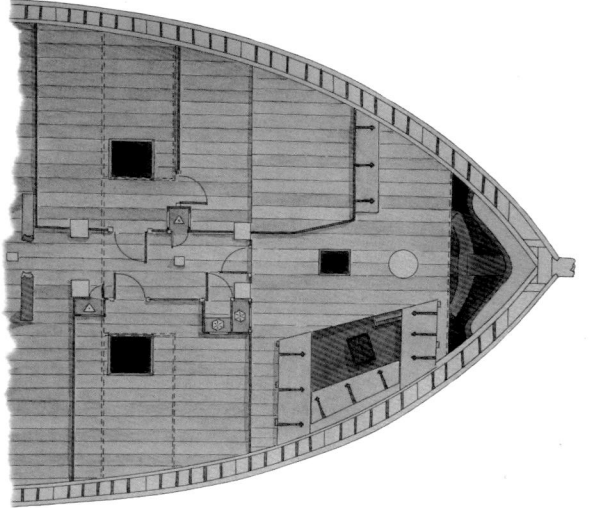

Fig. 2. Reconstruction of the forward platform of an English 70-gun warship, as arranged c.1690. The gunner's storeroom with cartridge chests is at right, with sunken filling room to starboard. The *London*'s platform, though without a light room, would have been similar. Copyright drawing by Richard Endsor.

outside the compartment. The one certainty is that there was ample opportunity for fatal carelessness.

Most shipboard powder disasters in the seventeenth century resulted from some ignorance or indiscipline, and the *London* may have been vulnerable in these respects. The ship had been in commission since June 1664 as flagship of Admiral Edward Montagu, Earl of Sandwich, commanding a fleet in the Channel.[8] When open hostilities with the Dutch broke out in December 1664,

Sandwich was designated as third-in-command of the huge 130-ship fleet to be set out for the next year's summer campaign and was assigned a larger flagship, the *Royal Prince*, which was fitting out at Chatham for the first time since her rebuilding in 1663. On 7 February Sandwich, having sailed the fleet to the Downs, left the *London* and dispatched her to Chatham for routine maintenance, graving, and replacement of ballast, to prepare her for her new commander, Vice-Admiral Lawson. Like Lawson, the Earl of Sandwich manned his ships as far as possible with his own personally loyal officers and seamen. Accordingly, Sir William Coventry, Navy Commissioner and secretary of the Duke of York, the Lord High Admiral, informed Pepys on 20 February that the *London*'s men 'are to be turned into the *Prince*, either by ticket or book'.[9] The phrase 'or book' refers to the pay book and implies that the whole ship's company was to go. Sandwich's flag captain in the *London*, Jonas Poole, was given another ship but the lieutenant, John Steward, and gunner Gabriel Walters went with Sandwich.[10] The other warrant officers were undoubtedly transferred as well. The

Fig. 3. Sir John Lawson, by Lely. Courtesy National Maritime Museum, London. Greenwich Hospital Collection.

emptied *London* then received an entirely new crew made up of Lawson's retainers supplemented by an unknown number of pressed men.

The change of crews took place while the ship was briefly drydocked for graving between 20 and 23 February, with Lawson's people coming aboard when she was refloated. The new gunner, Richard Hodges, had been selected in January and was already aboard,[11] but the intended lieutenant (probably Phillip Carteret) was not. The identity of the warrant master (assuming that one was aboard) has not come to light, so it is not clear who was in charge when the ship sailed. But there can be no doubt that the *London* had a new crew unfamiliar with the ship, with no one present more senior than a warrant officer. It must be acknowledged that there were many cases during this period in which warrant officers in battle performed heroically and competently when command of ships devolved upon them due to casualties among the commissioned officers; but in day-to-day activities, some were corrupt and not to be trusted out of sight of authority.

Samuel Pepys, in recording the disaster in his diary, was appalled by the loss of so many men and women. But Pepys, as a naval administrator, was also distressed that the ship had gone down 'with 80 pieces of brass ordinance'. He was wrong about the number of guns, but right that they were all brass. Cannons of bronze, or brass as it was called at the time, were preferred partly because of their handsome appearance, but also because they were more trustworthy than their iron counterparts. They were much less likely than iron weapons to burst without giving fair warning by bulging out of shape. But brass was expensive; about eight times the cost of iron. Once founders learned to manufacture reliable iron cannon, in the early seventeenth century, economy dictated that brass guns became increasingly less common except in the very largest sizes, for which iron-casting technology remained inadequate until the late 1660s. In the mid-seventeenth century brass guns of all calibres, even very old ones, were prized weapons. In March 1664[5] several English men-of-war had partial batteries of brass and quite a few had one or two pairs. But only four ships were armed entirely with brass: three first-rates, and the *London*.[12]

Considering the value of these weapons, it should not be surprising that the Board of Ordnance deployed considerable resources to recover guns after the catastrophe. If brass cannon were valuable then, they are still so today if only because of their rarity: worn out or broken brass guns, unlike those of iron, were not discarded but melted down for re-casting. Consequently, surviving seventeenth-century brass ordnance is for the most part limited to what has been recovered from the seafloor. As has been widely publicized, the remains of the *London* have received intensive archaeological attention in recent years and continuing through the present, the guns being objects of particular interest. The archaeologists, however, have been at something of a disadvantage in not knowing exactly what guns the *London* had aboard when she went down, or how many have already been raised over the course of three-and-a-half centuries. The remainder of this article will supply some of these wants, plus details of the ship and her brief but emphatically punctuated career. Along the way, the *London* will also serve as a revealing illustration of the political, economic, tactical, and technological factors involved in arming big ships of the mid-seventeenth century, and the special complications these matters presented during Protectorate and early Restoration times. In these periods a very few warships (mainly great first-rates) varied little in armament over their careers. But the *London* was far more typical in that her ordnance changed in both number and calibres with every commission, depending on what the Ordnance Office happened to have on hand at a given moment. As will be seen, the *London* at times also carried some rather unusual guns, with unusual histories.

'A Lusty Ship'

On 3 July 1654, the Council of State under Lord Protector Oliver Cromwell ordered the Admiralty Committee to undertake construction of 'four second-rates at such places as they think best'.[13] A month later the Admiralty directed the Navy Board 'to order the master shipwrights of Chatham, Woolwich, Deptford, and Portsmouth to prepare a draft or model of a second-rate ship to carry 60 guns'.[14] But the Navy was already in the midst of a tremendous shipbuilding programme undertaken during the just-concluded First Anglo-Dutch War, and finances were none too healthy. The Portsmouth second-rate never materialized, and the one at Woolwich was deferred, eventually emerging as the *Richard* in 1658. The other two were built with reasonable dispatch as planned. For the vessel that became the *London*, it is recorded that the master shipwright at Chatham, Captain John Taylor, submitted a 'platt', or draft, rather than a model (which, incidentally, meant in this context not a miniature ship, but a detailed list of dimensions and drafting data), to the Admiralty for approval.[15] When Taylor began construction is not recorded, but in an item dated 30 July 1656, the newspaper *Mercurius Politicus* reported that the Admiralty Commissioners, on hand at Chatham, had 'launched a lusty ship of the second rate named the *London*, carrying 60 great Guns'.[16] Her sister ship *Dunbar* (renamed *Henry* at the Restoration), was built at Deptford by master shipwright Manley Callis and launched in April 1656.[17]

Captain Taylor designed the *London* for a keel of 123 ft 6 in and an external beam of 39 ft 8 in, resulting in a burden of 1,035 tons; but as in almost all big ships of the seventeenth century, the beam 'fell out' slightly during construction. As completed, the breadth measured 40 ft 0 in, giving a burden of 1,051 tons. The depth in hold (plank to plank) was 16 ft 6 in and the designed draught (probably exceeded in service) was 18 ft 0 in.[18] Although the gundeck length is not recorded, 153 ft would be a reasonable estimate. In 1656 only three English warships, all first-rates, were larger.

The *London* and *Dunbar* were the first new second-rates to have been launched for some 22 years. Though much greater in dimensions than the earlier versions, both followed the same basic design as the earlier vessels. They had three complete flush decks with an entry port on the larboard side on the second deck. They had a short quarterdeck extending only a few feet forward of the mizzenmast and a small, almost vestigial, poop. In March 1655[6], with the ships nearly complete, the Admiralty Commissioners decided to order the shipwrights to add forecastles, which at the time were much in vogue owing to a perceived value

Fig. 4. The *Henry* (ex-*Dunbar*), by Willem Van de Velde the Elder c. 1661. Courtesy National Maritime Museum, London.

for defence against boarding. Captain Taylor, however, objected vociferously that this would 'misshape his ship', the decks forward having already been laid at an ample height.[19] The Commissioners accordingly conceded that the *London* and *Dunbar* could remain without a forecastle, though they insisted that the deferred second-rate (the *Richard*, still in the planning stage) be designed for a forecastle.[20]

Like all other second-rates built before 1658, the *London* and *Dunbar* were arranged internally as though they were two-deckers. The great cabin was on the second deck; the coach, or officers' dining cabin, and the roundhouse with its cabins for the lieutenant and the warrant master, were on the upper deck beneath the quarterdeck.[21] When Pepys visited the *London* in 1660, he noted that she 'hath a state-room much bigger than the *Naseby* – but not so rich'.[22] Although the *Naseby*, a first-rate, was the larger ship, her great cabin was on the upper deck and thus narrower than the *London*'s second-deck cabin because of the tumblehome. The *London*'s arrangement was not followed in subsequent second-rates, the space on the second deck being needed to accommodate the ever-growing complements.

The cookroom, with its brickwork and pair of copper 'furnaces', was preferentially 'in hold' in the 1650s. The hold at that time meant everything beneath the gundeck including partial decks, or platforms. Big ships had a platform in hold aft (the cockpit) extending forward from the mizzen with cabins and storerooms, and another platform forward as previously mentioned. Contrary to later practice, these platforms were not connected by a continuous 'orlop deck'. Instead, a ship of the size of the *London* had a row of perhaps ten 'false beams' positioned for strength six or seven feet beneath the gundeck. These were decked over only near the vessel's sides to form cable tiers, and were otherwise unplanked – except for the cookroom which was laid on the false beams in the forward hold just before the main hatch. In new ships of the mid-1660s the cookroom was re-positioned to the second deck abaft the foremast as one of several changes in arrangement to improve stowage capacity in the hold; a complete orlop followed a few years afterwards.[23]

Captain Taylor and the designer of the *Dunbar*, Manley Callis, planned for slightly more guns (64) than originally specified. Like most other three-deckers of the smaller class from this period, the ships had only two complete tiers of gunports, thirteen in the lower, twelve in the second. On the third (upper) deck the designers made provision for three guns on each side forward and five (later seven) on each side aft. The waist, where the gunwale dropped to only about a foot from the deck, was left unarmed. There was space for additional guns on the quarterdeck, but this was unarmed at first. The stern ports (four for the gundeck and two each for the second and third decks) were intended to share their guns with the adjacent broadside ports, the pieces being shifted as needed. The same was true of the bow chase ports in the forward bulkhead (two each on the middle and upper decks). The foremost gundeck port on each side, from which a traversed gun could bear forward, was also originally planned to share a gun with its neighbour because otherwise the recoil paths would have conflicted; but this economy was later ignored.

'For the Honner of ye Nation'

Arming big ships was no easy matter in Protectorate times. The introduction of the line of battle in the preceding Dutch War led to a desire to equip the fleet as far as possible with 'home-bored' or 'fortified' guns (rather than the previously ubiquitous short-range 'drakes') which could withstand heavy charges and thereby permit the stand-off effective ranges preferred for battle-line tactics. This presented no particular difficulties with medium-calibre weapons (18-pounder 'culverins' and 9-pounder 'demi-culverins'), but for vessels of the *London*'s size the gundeck would be expected to carry 32-pounder 'demi-cannon' and perhaps – or so the sea-officers hoped – even the huge 42-pounder 'cannon-of-seven' (so-called for its 7-inch bore) of which a fortified version had just been introduced. In the 1650s the casting of home-bored demi-cannon and cannon-of-seven was expensive, time-consuming, and technologically challenging. Few foundries were capable of undertaking it.

On 20 March 1655[6], with the *London* nearing completion, Taylor advised the Admiralty that he had designed his ship for a powerful armament, and to give 'gloree and Creditt for the nation' she should be armed with the 'greatest proportion that can be borne for forraine service upon such a body'. He continued, 'the proportion I meane for shipps of warre well Built is ten tunns of ordnance to every hundred tunns the ship ariseth to in burthen, calculated according to ye known rule'. To achieve this proportion he recommended that the guns 'be of the severall natures and weights as followeth':[24]

To be Planted one ye lower tyre   *cwt*
8 Cannons of 7 home Board Bras or Iron   poiz   54
14 Demy Canon vizt of Bras 2 of good length   54
        of Bras or Iron 12 home Board 45
2 Cullⁿ [culverins] of Bras length 12 feet fortifyd   50

Second tyre
16 Cullⁿ of Bras or Iron home Board   28
6 Demy Cullⁿ [demi-culverins] of Bras
   for ye Cabin etc   24
2 Demy Cutts [demi-culverin cutts] of Bras
   for ye gallerys   12

One ye upper Deck
2 Demy Cullⁿ upon ye fore castle:
   Bras 11 feet long   28
8 Demy Cullⁿ for ye Coach & upper Deck all Bras 22
6 Demy Cullⁿ Cutts Bras for Cuddy & Roundhouse   7

In all 64 Gunns: to poiz as above said 103 tunns & a halfe

The two 12ft culverins and the pair of 54cwt demi-cannon 'of good length' were the gundeck stern chasers, which had to be long enough to reach across the wing transom timber so the muzzles would extend clear of the gunports. Similarly, the 11ft demi-culverins forward on the upper deck may have needed extra length to cross the tails of the catheads; these could run either fore and aft or athwartships, but would interfere with the guns on one bearing or the other. It was also assumed (wrongly as was much later discovered) that long guns had inherently great range. At the other extreme, Taylor allowed short (5 to 7ft) demi-culverin 'cutts' for the after positions on the middle and upper decks. This was purely for the comfort of the officers, as longer guns would have taken up inconveniently more space in their cabins. In the body of his list, the shipwright specified cannon-of-seven, demi-cannon, and culverins 'of bras or iron'; but at the bottom he added a postscript: 'For the honner of ye nation I shall be sorry if such a ship must saile without all Bras ordnance'.

The Officers of the Ordnance, after consultations with Taylor and Chatham Dockyard Commissioner Peter Pett, sent their own recommendations for the ship's armament to the Admiralty and Navy Board on 6 May. They repeated Taylor's scheme with only trivial alterations: cannon-of-seven to be 53 cwt rather than 54, and the smallest demi-culverin cutts to be '7 or 8 cwt' rather than 7. But the Ordnance Officers did not offer a preference for brass or iron. On 17 July, with the ship ready for launching, the Admiralty and Navy Board jointly forwarded the Ordnance Office gun list to Taylor and Pett requesting their final opinions 'for preventing any alterations which hereafter may happen to be made to the delay and prejudice of the service'.[25] Two days later, the Council of State informed the Admiralty that the ship was to be called the *London*, no doubt reflecting the Lord Protector's appreciation of how important the City's support was for his political ascendancy.[26] Partly for this reason, but also because the ship would normally be employed as a flagship, brass ordnance was indeed assigned. With the government's parlous financial condition, however, there was no possibility of manufacturing 64 new brass guns; at the prevailing rate of £7½ per cwt, this would have cost over £15,000 – more than the ship. But brass culverins and demi-culverins were available from existing stores: old but well made English weapons, and many taken from various foreign enemies in recent wars. Brass demi-cannon could be obtained at a bearable cost by supplying the founders with obsolete guns (themselves in short supply) to be melted down for re-casting, and this procedure was probably used by the Ordnance Board's usual contract gunfounder, George Browne, to cast the *London*'s 8½ft, 45cwt demi-cannon (a special pattern not afterwards repeated).

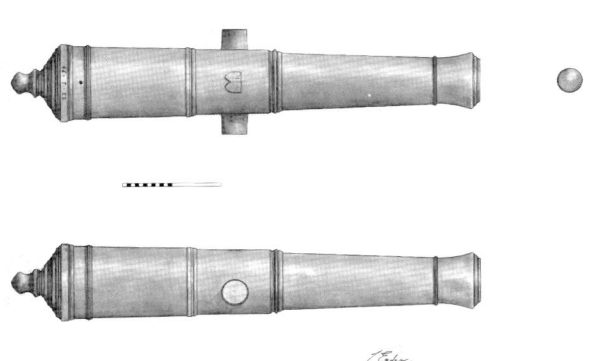

Fig. 5. Brass demi-cannon cast for the *London* by George Browne, recovered 2007. Copyright drawing by Richard Endsor.

Larger guns, however, were more troublesome. There is no production record of the intended pair of long 54cwt demi-cannon chasers, and the eight 53cwt cannon-of-seven (they would have been 9ft long) never appeared. Such pieces, if cast in brass, were too expensive for Cromwell's government to expend for a second rate. In all of Commonwealth and Protectorate times, and indeed through the first four years of Charles II's reign, the Ordnance Office acquired only 27 brass home-bored cannon-of-seven, all of which went to first-rates. Unfortunately, the manufacture of iron cannon-of-seven was still regarded as impractical. Founders feared that the required temperature of such large masses of molten iron could not be maintained during the pour for long enough to fill the moulds.[27] The result of these difficulties was that when it was decided to commission the *London* for the sea-campaign of 1657 against the Spaniards, she was initially assigned 64 brass guns of much lighter composition than previously planned:[28]

Lower Deck   12 demi-cannon
                     12 culverins
Middle Deck   12 culverins
                     12 demi-culverins
Upper Deck   16 demi-culverins

She would be able to fire only 1,068 lbs of shot as opposed to 1,324 lbs as intended.

Henry Quintyn, Gunfounder

All concerned with the *London*'s initial armament must have been disappointed with her lower tier. But while the ship was fitting out, an unexpected possible improvement presented itself. A year earlier, in April 1656, the Admiralty had received a petition from 'Henry Quintyn, Gunfounder'. Quintyn explained

That by his Industry and Diligent observation in the business of Gunfounding for severall yeares past and by many chargeable Experiments of M. Quintyn, [he] hath at length attained to the Founding of such Ordnance of Iron as were never done before in this Nation, if in the world.

He requested permission 'to make and deliver Cannon of 7 inches Diameter of Iron homeboared of 8½ foote in length or 9 foote if desired'.[29]

Quintyn, an accountant by training, was the former business manager for George Browne's father John Browne (Gunfounder to the King) in the 1640s, and in the early 1650s he was a partner of gunfounder Thomas Foley in a foundry manufacturing iron cauldrons. Foley was separately a partner of George Browne in the latter's gunfounding operations.[30] At the time, Foley and Browne enjoyed a virtual monopoly in supplying iron guns to the Ordnance Office, and Quintyn's petition went nowhere. But with the principal suppliers unwilling to attempt iron cannon-of-seven, and brass versions being unaffordable, Quintyn tried again. On 24 February 1656[7], he wrote to the Admiralty Commissioners:

Having heard that the London Friget is ordered to be equipt for the seas, And being sensible that out of zeale & affection to serve his Highness & my native Country, at greate Charge have accomplished the making of such Cannon of 7 of Iron homeboared as I shall willingly submit to ye proofe of brasse Guns of like Waite, and am so confident they are soe much better for service as that they will not heate so soone & being hot will sooner Coole. Their Waite is about 48 cwt apiece & 8½ foote long. Now my humble request is that you would consider whether it may be for ye Advantage of ye said vessell to carry some of these Guns & if so what number may bee thought fitting.

Quintyn concluded with an unprecedented proposition: he would deliver the guns promptly, 'and if not well liked after 6 monthes experience I shall take them againe, otherwise to be paid for them after the rate formerly given for ye best sorte of Iron Ordnance'.[31]

The Admiralty passed Quintyn's offer to the Ordnance Office. There, it was undoubtedly recognised that he had reduced the mass of the castings to a manageable size by designing short, lightweight weapons which (like cutts) would have a ferocious recoil. The offer was nevertheless too good to refuse, and accordingly eight cannon-of-seven (the number originally planned) were received into store on 21 March. Six of the guns were delivered by Richard Sholbury and two by a woman, Amy Shipps, both presumably managers of Quintyn's foundries.[32] One of these was on the upper Medway at Snodland, where Quintyn is recorded as having manufactured heavy iron mortars for the Ordnance Office, also in 1657.[33] Then, apparently on a different date, a *ninth* cannon-of-seven was delivered and also assigned to the *London*. Why this came about is unknown; perhaps one of the new brass demi-cannon failed its proofs and needed replacement. In any event, the ship went to sea with odd numbers but quite respectable calibres, including a lower tier with iron cannon-of-seven, brass demi-cannon, and brass culverin chasers.

For the *London*'s first commission, which began in June 1657, she was employed continuously either as the flagship in the Downs or at sea in nearby waters in support of the allied French and English land campaign against the Spanish and Royalists in Flanders. Admiral Sir Richard Stayner commanded her off and on for much of the commission, relieved at times by Captain William Whitehorne, and for a period in September and October 1657 by the General-at-Sea Edward Montagu, the future Earl of Sandwich.[34] It was during Montagu's tenure that the Officers of the Ordnance, having noticed that the six-month trial period for Quintyn's nine cannon-of-seven was about to expire, wrote asking the Admiralty Commissioners how the guns had performed.[35] The Admiralty Commissioners responded that a survey ordered by General Montagu had found the guns to be 'mushole [muzzle] heavy and thereby rendered unfit for the service'. The consequence of muzzle-heaviness (that is, the trunnions placed too near the breech) was a violent recoil. Since recoil troubles with Quintyn's guns were predictable due to their relatively light weight, the diagnosis of the surveyors is suspect. Nevertheless, the Admiralty's reply included a warrant directing that the guns be returned to Quintyn 'upon the comeing in of the said ship *London*'.[36] As it turned out, this intention was frustrated in that the *London* did not soon come in. She remained in commission as flagship of the 'winter guard', still based in the Downs, and then took part in the blockade which assisted in the siege of Dunkirk. Only after the fall of Dunkirk in June 1658 was the *London* sent into the Medway; she was paid off at Chatham in July.[37] By this time the Ordnance Officers had decided to retain Quintyn's guns despite their defects, and they had a considerable history ahead of them. The *London* was next commissioned for Montagu's expedition to the Baltic in 1659, which was intended to secure English access to that sea through a force-backed mediation of the bitter war between Sweden and Dutch-assisted Denmark. The ship again served as flagship for Stayner, the third-in-command.[38] With the mission unfulfilled, political turbulence in England caused Montagu to sail the fleet home in late August, and the *London* was paid off soon afterwards.[39] The ship as before bore 64 guns during the Baltic cruise, though the 'mix' of calibres might have differed.[40]

Henry Quintyn died in 1658.[41] His guns were among 16 iron cannon-of-seven possessed by the Ordnance Office in May 1664, and although briefly considered for reassignment to the *London* (as will be related shortly), this was not done. Instead, these 16 weapons served in the three great battles of the Second Anglo-Dutch War of 1664-1667 aboard the newer second-rates *Royal Katherine* and *Royal Oak* (ten in the former, six in the latter).[42] Later assignments at sea are unknown (documentation is weak for armament details in the Third Anglo-Dutch War), but five of Quintyn's cannon-of-seven were still on hand as spares at Woolwich as late as 1699.[43]

Restoration, Rest, and War

After a few months in ordinary in the autumn and winter of 1659-1660, the *London* went to sea again in the spring of 1660 under Lawson as Vice-Admiral of the fleet that bore Charles II to his restoration. The ship on that voyage supplied transportation for the King's brother, Lord High Admiral James, Duke of York. From this cruise on, the numbers of guns of each calibre aboard the *London* is documented. The most important of several archival sources is a tabular journal of issues and receipts of brass ordnance (and carriages and ammunition) maintained by the Ordnance Office branch at Chatham between October 1660 and early April 1665.[44]

In the autumn of 1660, with the other three-deckers laid up, the *London* stayed out under Lawson as flagship of the winter guard, during which she ferried Queen Henrietta Maria and Princess Henrietta to France.[45] The ship finally entered the Medway in April 1661 to be paid off, disarmed, and placed in ordinary. The guns taken out at that time, and which she had presumably carried throughout the commission, numbered 66, all brass:[46]

24 demi-cannon
24 culverins
16 demi-culverins
2 sakers (5¼-pounders)

The *London* spent the next three years moored unarmed off Chatham. Ships in ordinary inevitably deteriorated, and by August 1663 she had become leaky enough to require docking.[47] Possibly at this time (but certainly after 1660), shipwrights applied a 6-inch-thick girdling to the *London*, increasing her beam to 41 feet and her burden to 1,104 tons.[48] This improved her stability enough to permit an increase in armament. When the Navy Board and the Ordnance Officers negotiated a new ordnance establishment for the fleet in May 1664, the *London* was assigned 76 guns of impressive composition:[49]

12 cannon-of-seven
14 demi-cannon
26 culverins
4 demi-culverin cutts
20 sakers

In this establishment the specified armament of the largest vessels was curiously old-fashioned in that more guns were assigned than there were broadside ports; that is, precedence was given to the chase ports at the expense of the broadside. It appears that all four of the *London*'s gundeck stern chase ports would have been filled (with

Fig. 6. The *London* in 1660, by Willem Van de Velde the Elder. The Irish saltire in the Union flag appears in many drawings from 1660 and is apparently accurate for that time. Courtesy Atlas Van Stolk, Rotterdam.

the two longest demi-cannon and two long culverins), and two of these weapons would not have had access to the broadside ports. This condition, in which ships had been preferentially armed to emphasize all-around firepower for mêlée action before 1653, perhaps reflected the predominance on the Navy Board of shipwrights, elderly pre-battle-line sea-officers, and civilians with little understanding of the altered tactical needs. At the time, only Commissioner (and Admiral) Sir William Penn would have known better from battle-line experience, but he was fully occupied at the time in fitting out the fleet.[50] The errors were eventually corrected, probably on the insistence of the ships' captains, when the guns were actually issued.

The ordnance establishment of 1664 was adopted in expectation of the eventual mobilization of the fleet for war with the Dutch, intended to break their dominance of world trade. For this, Charles II's government had already initiated deliberate provocations in colonial areas. The whole establishment, however, was in fact hardly more than a wish list. An inventory of the guns on hand at the time, appended by the Ordnance Officers, shows that only 43 cannon-of-seven were actually available out of 108 specified, and only 392 demi-cannon out of 622 specified. The establishment was agreed upon just as the *London* was being fitted out as the flagship of a squadron in the Channel as the first contingent in the anticipated war fleet, and the shortage of heavy guns caused complications in her armament. Of the 43 fortified cannon-of-seven in store, 27 were brass guns reserved for first-rates. But there were still the 16 iron versions. The Ordnance Officers accordingly assigned her a lower tier of iron weapons which almost certainly included at least some of the cannon-of-seven originally made for her by Henry Quintyn. But on 25 May, Chatham Dockyard Commissioner Peter Pett wrote to his Navy Board colleague Samuel Pepys that the *London*, the only flagship planned for commissioning that summer, should have all-brass ordnance as a matter of prestige, 'but it seemes they have them not'.[51] Pepys undoubtedly shared Pett's disappointment, particularly when it became known a few days later that the Admiral aboard the *London* would be Pepys's cousin and patron, the Earl of Sandwich. An objection was lodged with the Officers of the Ordnance, who then faced a last-minute scramble to come up with something acceptable. They noticed that although sufficient brass cannon-of-seven and demi-cannon were unavailable, the inventory included 16 brass 24-pounders (not yet a standard English calibre), of which eight were unassigned. Since the appearance of the flagship's guns was obviously more important to Pepys and Pett than their firepower, these eight weapons became part of an all-brass outfit of 74 guns which was issued to the *London* on 7 June:[52]

16 demi-cannon
8 24-pounders
26 culverins
24 demi-culverins

Two of the culverins were presumably long stern-chase pieces for the lower deck, and four of the demi-culverins were cutts for the quarterdeck.

Fig. 7. Edward Montagu, 1st Earl of Sandwich, by Lely c.1658-59. Courtesy National Maritime Museum, London. Greenwich Hospital Collection.

On 20 June the Earl of Sandwich hoisted his flag in the *London*. The ship sailed with her squadron in July and remained in continuous commission for the next eight months, sometimes at sea in the Channel but more often anchored at Spithead and later in the Downs.[53] On 6 February 1664[5], while the fleet was in the Downs, Sandwich shifted his flag to the *Revenge* and sent the *London* to Chatham for the maintenance, graving, and change of command noted earlier. When she was ready for dry-docking on 20 February, her guns were taken out, and they and their carriages were entered separately on the 'receipts' pages of the Chatham ordnance journal. When the ship had been refloated, the same guns and carriages were re-embarked and entered on the 'issues' pages of the journal on 23 February. The armament was identical to that issued in the previous June, except that two brass 'minions' (4-pounders) had been acquired from an unknown source (perhaps Portsmouth) during the cruise. There were 76 guns in all:[54]

16 demi-cannon
8 24-pounders
26 culverins
24 demi-culverins
2 minions

As shown by the gunport arrangement in the drawings of the *London* by marine artist Willem Van de Velde the Elder, 26 guns including two long culverins would have been placed on the gundeck, 24 culverins on the middle deck, 20 demi-culverins on the upper deck (six forward, 14 aft), and on the quarterdeck four demi-culverins (cutts) and the two minions. These guns were listed just twelve days before the *London* blew up, so it was in this configuration that the ship sailed to her doom.

Salvage

In the aftermath of the disaster to the *London*, once the survivors had been rescued, the highest priority became the recovery of the brass guns. Tenders assigned to nearby large warships came alongside the wreck to take off the six guns on the quarterdeck, which was above water at low tide. On 9 March, two days after the explosion, Navy Board Commissioner Sir William Coventry could inform the Ordnance Officers that five of the guns were safely aboard the *Royal Katherine* and another aboard the *St George*.[55] The *Royal Katherine* had taken in four demi-culverin cutts and one minion.[56] The gun aboard the *St George*, though undocumented, was undoubtedly the second minion. On learning that these weapons had been received back into store, the Earl of Sandwich lost no time in writing to Sir Thomas Chicheley (de facto – and later actual – Master of the Ordnance) requesting that the four brass demi-culverin cutts be sent aboard his new flagship, the first rate *Royal Prince*, as 'I doe very much want them to compleat her upper tier'. This was so ordered.[57] Alas, these guns were lost again the next year when the *Royal Prince* herself went down on the Galloper Sand during the Four Days' Battle.

The Navy Board and Ordnance Office engaged several small vessels soon after the accident in hopes of salvaging not only the *London*'s guns, but the hull as well: the smack *John of Stroud* under John Bickford and the *Prosperous* hoy under John Marshall, both in March; and the *Sarah* pink under Lambert Wood soon after.[58] The most important salvage vessel was the *Harwich* hoy under Giles Bond, which was employed at the site for many months. Initially, little was accomplished; Captain Bond reported to Pepys on 15 May that he had recovered a small hawser and 19 pieces of cable, but no guns. He added, in a comment that modern archaeologists would appreciate, that 'the weather is so bad the divers have only been down three times'.[59] But two days earlier the Ordnance Office had acquired two large diving bells from one George Cave.[60] Over the next five months Bond's divers, using the bells, painstakingly (and no doubt harrowingly) hauled up 18 guns: four demi-cannon, seven culverins, and seven demi-culverins.[61] The approaching winter put an end to the operations, and all was quiet at the site for the next 14 years.

William Harrington, Inventor

In 1679 a seaman with an engineering bent named William Harrington contracted with the Ordnance Office to attempt the recovery of the *London*'s guns by a new method he had devised 'to be wrought without diving'.[62] On

28 February 1679[80] Harrington petitioned the Privy Council, 'Praying payment of £854, 5s for Weighing severall guns amongst w[ch] 7 of Brass & Copper out of the Old *London*'s Wreck'.[63] But as nearly all contractors of the period discovered, extracting payments from Charles II's government was difficult. On 8 December 1680, Harrington complained to the Privy Council that by contract he was due 'half of the clear value' of any guns recovered from the *London*. The guns, however, had been taken from Tower Wharf by Alderman Sir William Pritchard, 'who not only hindered the Pet[r] from his Reward but obstructs the saving the rest of the Guns of the said Ship'.[64] Pritchard was an important supplier of cordage and, reading between the lines, was perhaps one of Harrington's unpaid subcontractors. The Privy Council forwarded the complaint to the Ordnance Office for adjudication. The results are undocumented; but it appears that Harrington's claims were upheld, for in early 1682 he and two financial backers were granted a patent on their salvage 'engines', allowing them (as was subsequently ruled) full rights to any of the King's guns that they recovered.[65] For seven years thereafter there are no records of activity at the site and it may be that Harrington, a mariner, was at sea. But in 1689 he returned for another campaign, which resulted in the recovery of 'two great Brass and Copper Gunns out of his Majesty's Frigat the *London*', which in May 1690 were duly ruled to be his property under the patent.[66] Harrington must have assumed that the Ordnance Office would buy the guns; but they apparently declined to do so, having ceased all efforts to acquire brass guns. With no hope of recovering more than the scrap value of the metal, Harrington's salvage efforts came to an end.

Two aspects of Harrington's operations may have relevance today. First, an entry in the Privy Council Registers mentions that he recovered both brass *and iron* cannon,[67] and indeed, archaeologists have recently identified probable iron guns at the site. No iron weapons were part of the *London*'s armament, but broken or worn out iron ordnance was sometimes employed as ballast. The *London*'s ballast, replaced in late February 1664[5], included more than the usual shingle or stone locally available at Chatham, since it was shipped to her from somewhere in the Thames above Deptford, off which one of the six hoys carrying it was stopped and her crew illegally pressed.[68] Perhaps it was old iron guns from the Ordnance Office stores in the Tower.

Also possible is that the iron guns recovered by Harrington were deposited in or near the *London*'s wreck site during the battle at the Buoy of the Nore that took place on 26 July 1667. In this action, which was fought on the flood tide in an easterly wind, the English expended four fireships, each carrying two to six iron guns, in an unsuccessful attack from downstream on the blockading Dutch fleet. The fireships had been so inexpertly prepared that they took an inordinate amount of time to ignite, so the wind and current could easily have carried them to the *London*'s position just upstream before they burned out and sank.[69]

A second aspect of Harrington's work that may have noteworthy implications today involves the current condition of the wreck site, which is spread over a quite broad area. Originally, the after half of the hull was in one piece so that there was hope of recovering it as of October 1665.[70] Harrington's patented salvage method without using divers is nowhere described, but he must have trawled or mechanically clawed his way through the wreckage, which could have contributed to its dispersal beyond natural processes.

Modern Salvage Efforts

For some three centuries after Harrington's activities, salvers left the *London* undisturbed. In the early 1960s (probably 1961), the Port of London Authority's Wreck Raising Service under Captain G R Rees, while investigating a rediscovered nineteenth century wreck which had settled amid the *London*'s forgotten remains, were surprised when their wire sweep snagged a splendid seventeenth century brass cannon. This turned out to be a French-made demi-culverin cast in 1636.[71] The gun could have come into English hands from any of a number of French warships captured between 1650 and 1654; in September 1652 Blake's Commonwealth fleet took seven while they were attempting to raise the Spanish siege of Dunkirk in defiance of English warnings.[72] The gun is displayed at the Royal Armouries Museum at Fort Nelson, Portsmouth. Further activity occurred in 1980, when the Port of London Authority's salvage vessel *Yantlet* hauled up two guns from the site.[73] Descriptions of these weapons and what became of them are at present unknown.

In 1985, Chatham Historical Dockyard, seeking brass cannons for display, and learning from ordnance historian Charles Trollope that the *London* wreck would be a prime source, approached the Admiralty for assistance. The Royal Navy accordingly arranged for HMS *Sheraton*, a minesweeper, to conduct detection gear exercises over the site. The results could not be fully made public, but did indicate the presence of numerous guns.[74] Unfortunately, funds to exploit this knowledge were unavailable, and nothing more was accomplished at the time.

Only in the last five years have serious systematic exploration and recovery efforts begun, spurred by two previously unanticipated developments: first, the realisation that the enlargement of the shipping channel for a new container ship terminal (part of the London Gateway project), with dredging due to begin in 2008, would adversely affect the wreck area; and second, the discovery in 2007 that the site was already being worked by a private diving team whose activities, though legal, did not conform to accepted archaeological protocols. With enthusiastic public support, in 2008 the site was designated as a restricted area under the Protection of Wrecks Act of 1973. With the blessings of the Port of London Authority and the developer the shipping channel was re-routed away from the wreck, and unauthorized diving activity in the designated area was forbidden. Wessex Archaeology Limited, under commission from English Heritage, plus an appointed private licensee, carry out current work at the site.[75]

Considerable controversy exists concerning the private salvage activities that took place before the *London* gained protected status. During 2007 the leader of a salvage group which had been active at the site declared the recovery of five large brass guns to the Receiver of Wreck. Two of these, a Commonwealth demi-cannon and an unusual English 24-pounder, were immediately recognized as Crown property and accordingly appropriated. The salver acknowledged that these had come from the *London* wreck. But he reported that the other three guns, all Dutch 24-pounders, had come from an undisclosed site in the lower Thames estuary, in international waters and therefore the property of the finder. Many divers and archaeologists have treated this claim with scepticism.[76]

While the Dutch guns were temporarily in the custody of the Royal Armouries Museum at Fort Nelson for conservation, Charles Trollope noted that one of them bore unusual ferrous concretions which appeared to be identical to those on the two English weapons, suggesting rather strongly that these had been in the same environment within the same wreck. The absence of concretions on the other two Dutch guns is not evidence that they were from a different wreck, but merely that they were in a different micro-environment not closely exposed to iron objects; this would be expected for guns that were thrown clear by the explosion. Dutch gun expert Nico Brinck also visited Fort Nelson. He found that the Dutch guns dated from 1600, 1616, and 1617, all having been cast by known founders for the city of Amsterdam. They were elaborately decorated, and were identifiable as land service guns from the city's defences.

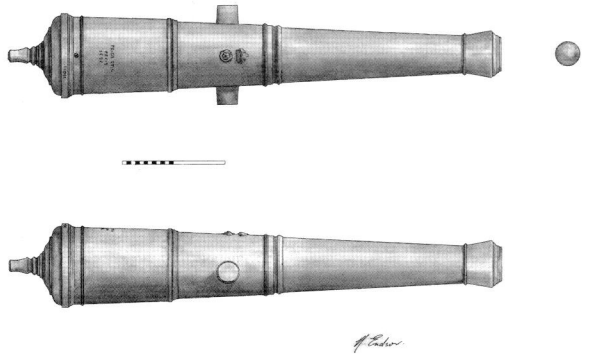

Fig. 8. Tudor-period 24-pounder from the *London*, originally cast as a culverin. Copyright drawing by Richard Endsor.

He also noted that tampions were in the muzzles, showing that the ship from which

Fig. 9. Fragment of a tampion found in the muzzle of a gun recovered from the *London* in 2007. Photo courtesy of Richard Endsor.

they came had been lost not in battle, but by accident.[77] Historians have so far identified no vessel armed with Dutch guns of this calibre and age as having been lost accidentally in the lower Thames estuary. Nevertheless, with no definite evidence of where the recoveries had occurred, and no detailed knowledge of the *London*'s armament, the Receiver of Wreck eventually released the guns to the finder, who sold them at auction, fortunately to a reputable collector.

In view of the controversy, the origins of the *London*'s brass 24-pounders have obvious relevance. As noted previously, the Ordnance Office had 16 on hand as of May 1664. One of these (recovered from the *London* in 2007) was originally a heavy English Tudor-period culverin which had been bored up to 24-pounder calibre, perhaps in the 1660s to fill out an even-numbered set. But 24-pounders were not manufactured in England until 1666, and those previously employed were mostly captured from the Dutch. Aside from the 16 available in 1664, three others found unfit for further service were melted down two years earlier – part of a group of seven which had been shipped off to the defences of Dunkirk in 1660, but brought back following its sale to France in 1662.[78] The melted weapons probably included two 24-pounder *kamerstukken* (approximate equivalents of light-bodied English drakes) which had been acquired from the Dutch ship *Wapen van Rotterdam* taken in 1652. These guns would have been useful as fortress weapons firing canister from enfilading positions, but were not liked by English sea-gunners. They do not appear in later

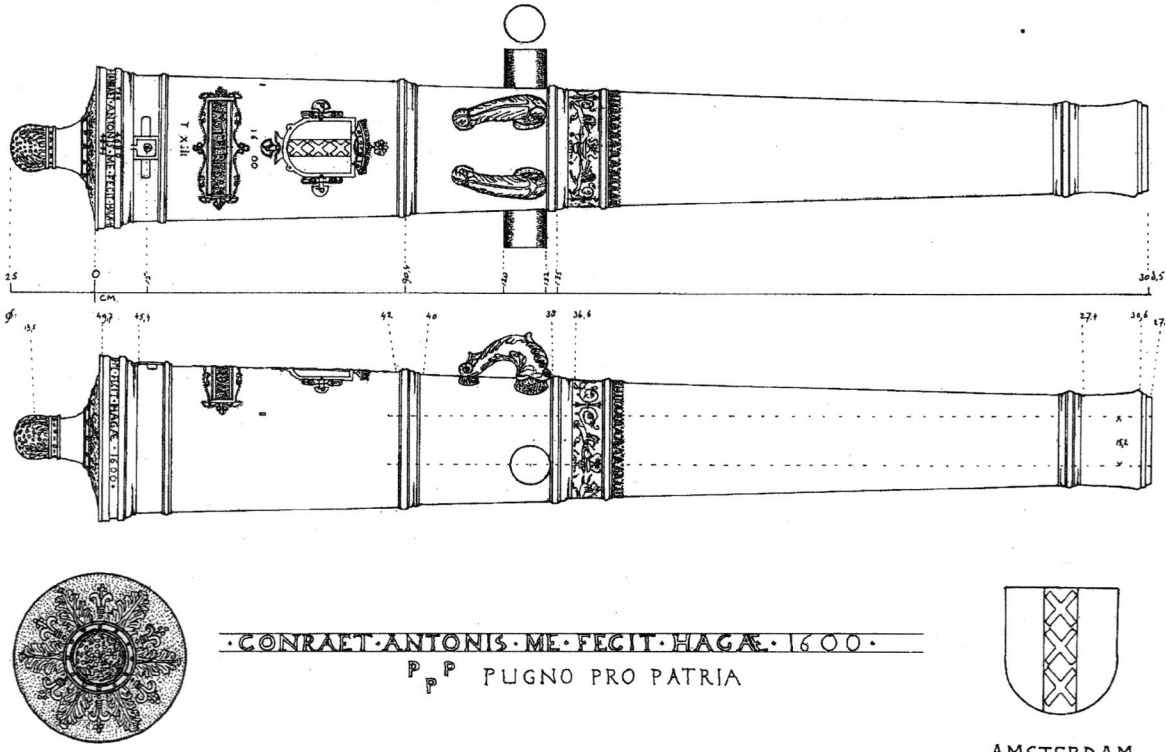

Fig. 10. Dutch 24-pounder cast 1600 for the Amsterdam defences. Drawn by Nico Brinck at Fort Nelson, Portsmouth, 2008; copyright reserved.

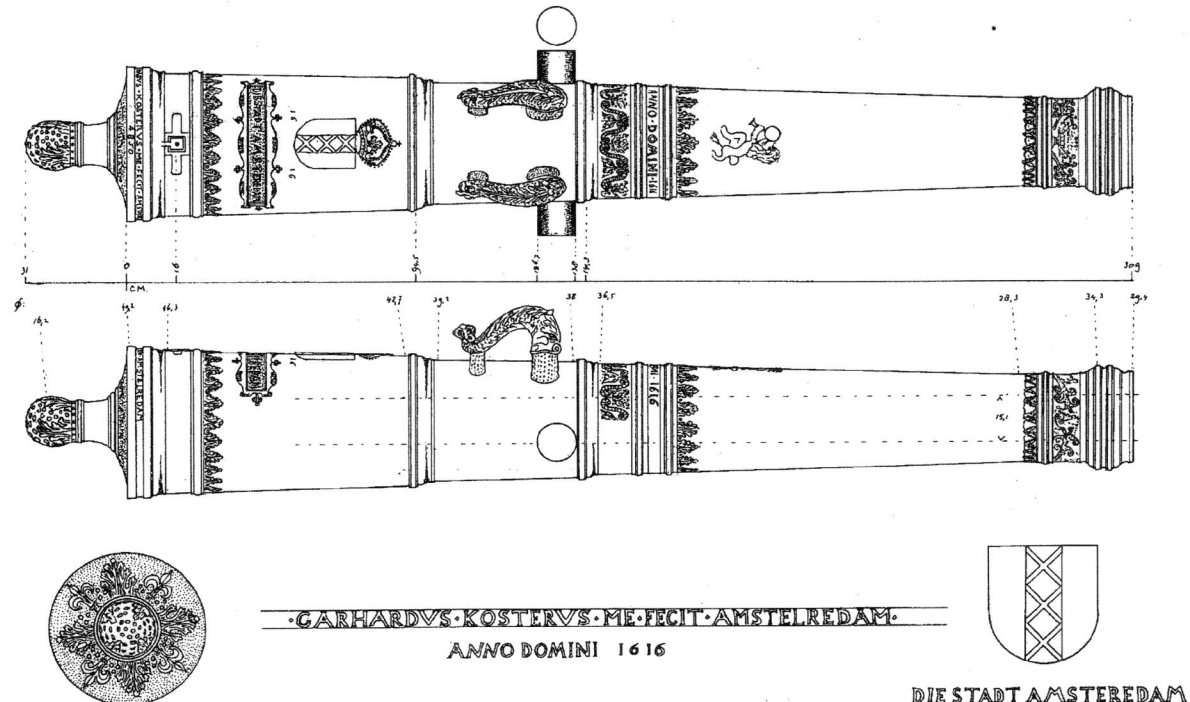

Fig. 11. Dutch 24-pounder cast 1616 for the Amsterdam defences. Drawn by Nico Brinck at Fort Nelson, Portsmouth. 2008; copyright reserved.

inventories. Deducting the two *kamerstukken* and the recovered re-bored English weapon leaves sixteen 24-pounders to be accounted for. Some are likely to have come from the large Dutch ship *Gewapende Ruijter* taken unawares while returning from Brazil at the start of the First Dutch War. Her guns were not Dutch, as she had been acquired as a prize off Brazil, though which prize is uncertain; possibly the Portuguese *Sao Bartolomeu*, but more likely the French *Villeroi* taken in 1650.[79] With greater certainty, Dutch records show that the other warships taken by the English in 1652-1654 included four with fortified brass 24-pounders aboard:

Prinses Roijaal Marie (two 24
 pounders), impounded at Dover, June 1652.
Groote Liefde (two 24-pounders), taken Battle of Portland, February 1653.
Elias (four 24-pounders), taken Battle of the Gabbard, June 1653.
Sint Mattheus (four 24-pounders), taken Battle of the Gabbard, June 1653.

The first named belonged to the Admiralty of the Maas (Rotterdam). But of greater interest are the *Groote Liefde*, *Elias*, and *Sint Mattheus* (carrying among them ten brass 24-pounders), all of which were Amsterdam directors' ships.[80]

Between 1631 and 1656 the warships of the Dutch navy were supplemented by large merchantmen hired, equipped, manned, and administered by town boards of directors completely independent of the five admiralties. For most of the period directors' ships were unimpressively armed, but during the First Anglo-Dutch War the Netherlands faced no land threat. Some towns accordingly employed their large defensive fortress cannons aboard the directors' ships. The directors of Amsterdam, the wealthiest city, hired some 35 merchantmen during the war and supplied ten of them with one or two pairs of brass 24-pounders. Of these ten ships, six survived the war to return their guns to the city, and one foundered in a storm off the Shetlands, but the three listed above were captured by the English and their guns happily received into store by the Ordnance Office.[81] Thus, it should come as no surprise that high-quality brass 24-pounders from the city of Amsterdam would have been aboard the *London*. On the other hand, it is hard to understand what other wreck could have been the source of the disputed guns.

This article enumerates the recovery of 41 of the *London*'s brass guns: 24 in 1665, nine in 1679-89, and probably eight in modern times. Assuming that the unknown guns brought up in 1980 were of brass and also that the three Dutch 24-pounders came from the *London*, there should be 35 brass guns still awaiting recovery, plus iron weapons used as ballast. If the guns are found, their salvage will in all likelihood depend on whether museums can be found, public or private, with the willingness to conserve and display them.

Conclusion

The disaster to the *London* in 1665 has presented historians and archaeologists with many questions. Some answers have been supplied here, such as the details of the ship's armament when she went down, and some additions to what was already known about the guns salvaged in the immediate aftermath. But uncertainties such as the causes of the explosion must remain open to speculation. Various situations can be imagined concerning the activities of her new crew at the time of the accident, but nothing can be proved. In the reconstruction of the history of the *London*'s armament, it must be stressed that the ever-changing numbers and calibres of the guns assigned to her were not at all unusual in the mid-seventeenth century; arming her sister ship was no less complicated. As explained here, many elements affected the guns assigned at any given time. Tactics dictated the kinds of weapons desired, such as the change in emphasis from short-range drakes to long-range fortified guns brought about by battle-line fighting after 1653. The prestige of the commander or even the ship's name could influence the choice of brass or iron guns. And finally, finances and technological limits, especially for the largest cannons of the *London*'s time, determined whether or not the Admiralty could have the guns it wanted. I (and the generous scholars who helped me) hope that this article may be of use to archaeologists in their continuing efforts to discover the wealth of additional information about the *London* that undoubtedly lies amid her remains on the bed of the Thames.

Appendix: Brass Guns Recovered From The *London*

The weights of English guns were normally inscribed in hundredweights (cwt) of 112 pounds, quarters of 28 pounds, and remainders in pounds. Some prize guns were weighed and inscribed in the English manner, but others retained their foreign weight inscriptions unless altered by re-boring or cutting. The weights serve as unofficial identification numbers in listings and inventories. Lengths are breech ring to muzzle.

Recovered 8-9 March 1664[5]
Demi-Culverin Cutts taken aboard the *Royal Katherine*
1. 21-1-10 Origin unknown
2. 14-0-14 Origin unknown
3. 13-3-10 Origin unknown
4. 11-2-14 Origin unknown

Minion taken aboard the *Royal Katherine*
5. 7-0-08 Origin unknown

Unknown gun taken aboard the *St George*
6. Probably a minion

Recovered by the *Harwich* hoy before 24 October 1665.
Dr Peter Le Fevre discovered the document recording the recovery of these guns, and Charles Trollope established the origins of many of them.[82] My own contribution involves only possible later re-issues and a few minor details.

Demi-Cannon: part of a set of 8' 6" guns cast by George Browne in 1656-1657. How many guns of this distinctive pattern Browne made is unclear. The *London* initially was to have had twelve, but her sister ship *Dunbar* also needed demi-cannon. Another from this series was recovered from the *London* in 2007; see no. 35 below.
7. 44-3-02
8. 44-0-11
9. 43-3-08
10. 42-3-14

Culverins
11. 39-0-27   Probably 8' 6" long. Cast by Henry Pitt 1611, and assigned to the *Anne Royal* in 1622.

12. 36-2-25 Cast by Samuel Owen 1596, assigned to the *Triumph* 1609, and in store in 1622 (the *Triumph* having been broken up).

13. 36-0-05 At least two guns of this weight. An 8' 6" weapon, caster unknown from pre-1593, was aboard the *Rainbow* in 1593 and later the *Reformation* and *Red Lion* in 1622. Another, cast by George Elkin in 1596, belonged to the *Triumph* in 1609, and was in store in 1622. Whichever it was, a gun of the same weight (and 8' 6" long) was still in use in 1699, assigned to the *Royal William*.[83]

14. 35-3-00 A gun of this weight, probably made pre-1593, was assigned to the *Red Lion* in 1622.

15. 35-2-15 Cast by George Elkin 1593, assigned to the *Red Lion* 1622.

16. 34-0-20 Origin unknown, perhaps foreign.

17. 34-0-12 Origin unknown, perhaps foreign.

Demi-Culverins

18. 29-1-01 Cast by Richard Phillips 1613, 8' 8" long.

19. 26-1-01 No record, possibly French like the gun recovered in 1961; see no. 34.

20. 26-0-00 At least two of this weight are known. One, caster unrecorded from pre-1593, 8' 0" long, was assigned to the *Merhonour* in 1622. A 9' 0" gun of this weight, possibly French, was assigned to the *Royal William* in 1699.[84]

21. 25-0-10 No record, possibly French.

22. 23-3-18 Caster unrecorded, 8' 6" long, assigned to the *Aid* 1593.

23. 23-0-16 Caster unrecorded pre-1593, assigned to the *Bonaventure* 1609.

24. 22-2-14 No record, perhaps French.

Recovered by William Harrington, 1679-89

25-31. Seven guns raised 1679-80. Valued by the salver at £854 5s. Based on the prices typically paid by the Ordnance Office for captured brass guns, Charles Trollope estimates the average weight of these guns at about 40 cwt, which implies that they included some demi-cannon. There is no record that any of these were re-issued.

32-33. Two unknown guns recovered 1689. No record of purchase or re-issue by the Ordnance Office.

Recovered by the Port of London Authority, 1961

34. Demi-Culverin, 23-3-00, length 9' 0". French, cast 1636, captured 1650-1654. Now in the Royal Armouries Museum, Fort Nelson, Portsmouth.

Recovered by the Port of London Authority, 1980

35-36. Two unknown guns raised by the salvage vessel *Yantlet*.

Recovered in 2007

37. Demi-Cannon, 46-3-22, length 8' 6". Cast by George Browne 1656-1657, part of a set made for the *London* and perhaps the *Dunbar*. Now in the Royal Armouries Museum, Fort Nelson, Portsmouth.

38. 24-pounder, weight not visible. Length 8' 6", cast by Peter Gill c.1595 as a culverin, but later (perhaps 1660s) re-bored as a 24-pounder. Now in the Royal Armouries Museum, Fort Nelson, Portsmouth.

39. 24-pounder. Inscribed weight 4,680 Amsterdam pounds (of 494 grams). Length 11 Malines feet (of 280 mm), or 10 English feet. Cast by Coenraet Antonis in The Hague in 1600 as a land service weapon for the defences of Amsterdam. The gun has lifting rings of dolphins with acanthus leaves. The barrel is decorated with friezes and rings of acanthus leaves. The first reinforce has a cartouche inscribed 'AMSTELREDAM', with the date 1600 and the arms of Amsterdam nearby. Currently in a private collection.

40. 24-pounder. Inscribed weight 4,850 Amsterdam pounds. Length 11 Malines feet or 10 English feet. Engraved with the name of the caster, Gerard Koster, the date 1616, the words 'DIE STADT AMSTEREDAM' and the arms of the city. Elaborately decorated, with dolphin lifting rings. A land-service gun with corniced muzzle rings, intended for the Amsterdam defences. Currently in a private collection.

41. 24-pounder. Cast by Gerard Koster and identical to no. 38 except for the year, 1617. Weight illegible due to concretions and erosion. Currently in a private collection.[85]

References

1. *The Diary of Samuel Pepys*, ed. R Latham and W Matthews, 11 volumes, Berkeley, 1970-1983, VI, 52, 8 March 1664[5].

2. *Calendar of State Papers, Domestic Series* [*CSPD*]: The Commonwealth, 1649-1660, ed. M A E Green (13 volumes, London, 1875-1886); Charles II, 1660-1685, ed. M A E Green, F H B Daniell, and F Bickley (28 volumes, London, 1860-1939), *1664-1665*, 249-250, 10/20 March 1664[5], Van Goch to States General.

3. *The Journal of Edward Montagu, First Earl of Sandwich*, ed. R C Anderson, Navy Records Society, London, 1929, 171.

4. *The Diary of John Evelyn*, ed. E S de Beer, London, 1959, 473, 9 March 1664[5]; and 475, 14 May 1665.

5. *CSPD 1664-1665*, 246-247, 9 March 1664[5], anonymous to Bennet.

6. The National Archives, Kew [TNA], State Papers [SP] 29/102, fo. 1, for the first known mention of a filling room (needed in merchantmen hired as men of war) in 1664, though the context shows that it had long been a standard feature.

7. TNA, War Office Papers [WO] 55/331, entry book, 13 March 1664[5].

8. *Sandwich Journal*, 143-167.

9. *CSPD 1664-1665*, 205, 20 February 1664[5].

10. TNA, Admiralty Papers [ADM] 10/15, commissioned officers 1660-1685, for Poole and Steward; TNA, WO55/331, entry book, 25 February 1664[5] for Walters.

11. TNA, ADM106/10, Navy Board in-letters, 19 January 1664[5].

12. TNA, WO55/1667, issues and receipts of guns and stores at Chatham, 1660-1665.

13. *CSPD 1654*, 241.

14. *Ibid.*, 536, 2 August 1654.

15. TNA, SP18/137, fo. 32, Taylor to Admiralty Commissioners. I am indebted to Dr Peter Le Fevre for discovering this important document, which contains far more valuable information (such as this detail) than is apparent from the *CSPD* summary.

16. *Mercurius Politicus*, no. 320, 31 July 1656. Thanks to Dr J D Davies for this reference.

17. *CSPD 1655-1656*, 524, Callis to Navy Commissioners; *Ibid.*, 526, Scott to Navy Commissioners.

18. The intended burden of 1,035 tons (from which the designed breadth is easily calculated) is implied in TNA, SP18/137, fo. 32 (see note 15). The 'as completed' burden as of 1660 is given as a slightly miscalculated 1,050 tons in Pepys's 'Register of Ships' in *A Descriptive Catalogue of the Naval Manuscripts in the Pepysian Library at Magdalene College, Cambridge*, ed. J R Tanner, 4 volumes, Navy Records Society, 1903-1922, I, 256; for depth in hold and designed draught, along with later altered beam and tonnage as of 1665, see *Ibid.*, 266.

19. TNA, SP18/136, fo. 305, report by Peter Pett, 16 March 1655[6].

20. *CSPD 1655-1656*, 572, 28 June 1656, Admiralty to Navy Commissioners; *CSPD 1656-1657*, 393, 14 July 1656, Navy Commissioners to Admiralty.

21. TNA, SP18/137, fo. 32, Taylor to Admiralty (see note 15).
22. Pepys, *Diary*, I, 114, 24 April 1660.

23. A specification for a big Commonwealth two-decker in TNA, SP46/136, fos 395-400, requiring a cookroom in hold, was used as the model for two new third-rates ordered in 1664-1665. Navy Commissioners made numerous alterations to bring it up to date, but the change in the cookroom position was a last-minute decision which only appeared in the final contract; see British Library, Additional MSS 9307, fos 41-46. For a description of the position of the cookroom in hold, see *A Treatise on Shipbuilding and a Treatise on Rigging Written about 1620-1625*, ed. W Salisbury and R C Anderson, Society for Nautical Research Occasional Publications, no. 6, London, 1958, 11-12.

24. TNA, SP18/137, fo. 32, Taylor to Admiralty (see note 15); format here is slightly altered from the original.

25. TNA, ADM2/1729, Admiralty out-letters, fo. 93v.

26. TNA, SP18/129, fo. 71, Pett to Admiralty, 16 July 1656 but included in the Council of State proceedings of 18 July; and *CSPD 1656-1657*, 30, Admiralty Commissioners to Navy Commissioners, 19 July, reporting the ship's name.

27. S B Bailey, *Prince Rupert's Patent Guns*, Royal Armouries Monograph 6, Leeds, 2000, 32, prints a letter from George Browne showing that this view still prevailed in September 1664 (citing a private collection, copy in the Kent Archives Office, Maidstone, TR 1295/69).

28. E Fraser, *The Londons of the British Fleet*, London, 1908, 47, citing an unspecified manuscript in the British Museum (which would now be in the British Library). Fraser, unaware of subsequent alterations, supposed that this was the armament the ship took to sea.

29. TNA, SP18/139, fo. 130.

30. B L C Johnson, 'The Foley Partnerships: The Iron Industry at the End of the Charcoal Era', in *Economic History Review*, New Series, Vol. 4, No. 3, 1952, 322-340, p. 327, n. 1.

31. TNA, SP18/162, fos 112-113.

32. TNA, WO51/2, bill book; and WO 49/91, debenture book, both 21 March 1656[7].

33. C Ffolkes, *The Gun-Founders of England*, Cambridge, 1935, 38.

34. TNA, SP18, numerous documents from May 1657 through July 1658 trace the ship's activities and changing commanders.

35. TNA, SP18/171, fo. 158, 26 Sept. 1657, listing the weights of the nine guns; copy in TNA, WO 47/4, Ordnance Office proceedings, fo. 58v.

36. TNA, SP18/171, fo. 160.

37. *CSPD 1658-1659*, 426, 1 July 1658, Taylor to Navy Commissioners; 428, 6 July 1658, Admiralty to Navy Commissioners; 428, 8 July 1658, Taylor to Navy Commissioner Bourne.

38. *Sandwich Journal*, xvii-xviii and 3-46.

39. TNA, SP18/216, fos 123-124, list of ships unpaid as of November 1659, the *London* omitted and thus already paid.

40. TNA, WO47/4, Ordnance Office proceedings, fo. 232.

41. *CSPD 1658-1659*, 113, proclamation by the Protector of 12 August 1658 relating to Quintyn's death and his replacement as a Commissioner of the Excise.

42. TNA, WO55/1652, gunners' charges and returns, 1666.

43. TNA, WO55/1736, fo. 48v, ordnance survey of c.1699, survey numbers 3429-3433; credit is due to Charles Trollope for noticing (from the weights) that these were Quintyn's guns.

44. TNA, WO55/1667, issues and receipts.

45. TNA, WO55/463, Ordnance Office warrants, 22 October 1660; *Sandwich Journal*, 82-86, abstract from Lawson's journal.

46. TNA, WO55/1667, issues and receipts; TNA, PRO 30/37/8, receipts of guns and stores, 12 April 1661, shows that no guns remained aboard.

47. *CSPD 1663-1664*, 235-236, 12 August 1663, Commissioner Peter Pett to Navy Commissioners.

48. Pepys, 'Register of Ships', in *Cat. of Pepysian MSS*, I, 266.

49. TNA, WO47/5, Ordnance Office proceedings, fos 121-122.

50. Pepys, *Diary*, V, 152, 18 May 1664.

51. TNA, SP29/98, fo. 219.

52. TNA, WO55/1667, issues and receipts; also listed in TNA, PRO 30/37/7, deliveries of guns and stores, 16 June 1664 as received by gunner Edward Curtis that day, which is probably when the last guns came aboard.

53. *Sandwich Journal*, 143-167. Sandwich briefly traveled overland to Whitehall in October to confer with the king, but the ship remained at Spithead.

54. TNA, WO55/1667, issues and receipts. A potential point of confusion: the *London* occasionally received many shot for rabonets (½-pounder swivels), but she had no such weapons. The shot were to be bagged for use in larger guns; see TNA, WO 55/331, entry book, 26 November 1664.

55. TNA, WO51/5, bill book, 9 March 1664[5].

56. TNA, PRO30/37/8, receipts of guns and stores, 17 March 1664[5], listed by weight.

57 TNA, WO51/5, bill book, 19 March 1664[5]; and TNA, WO 55/331, entry book, 23 March 1664[5].

58 TNA, WO51/5, bill book, for the first two, and TNA, SP29/149, fos 122-136, journal of the *Sarah*.

59 *CSPD 1664-1665*, 316, 15 May 1665, Bond to Pepys.

60 TNA, WO51/5, bill book, 13 May 1665.

61 TNA, SP29/135, fo. 71, Bond to Navy Board, 24 October 1665, listing calibres and weights. Many thanks to Dr Peter Le Fevre for his discovery of this important paper.

62 *CSPD 1680-1681*, 164, 12 February 1680[1], patent application.

63 TNA, Registers of the Privy Council [PC] 2/68, 372. I am indebted to Charles Trollope for directing me to this and the following PC entries.

64 TNA, PC2/69, 162.

65 *CSPD 1682*, 52-53, 1 February 1681[2]; TNA, PC2/69, 502.

66 TNA, PC2/73, 194, 22 July 1689 and 438, 15 May 1690.

67 TNA, PC2/69, 161.

68 *CSPD 1664-1665*, 205, 20 February 1664[5], Coventry to Pepys; 209, 20 February 1664[5], Captain Badiley to Navy Board; 216, 24 February 1664[5], Badiley to Pepys.

69 Bodleian Library, Oxford, Rawlinson MSS A.195, fo. 264, a detailed account. Secondary sources list five fireships expended, but one of these, the *Elizabeth & Mary*, survived; see TNA, ADM 10/15, commissioned officers 1660-1685, for John Votier.

70 *CSPD 1665-1666*, 13, 14 October 1665, Brounker, Mennes, and Berisford to Pepys.

71 R Jefferis and K McDonald, *The Wreck Hunters*, New York, 1966, 181-2.

72 *Letters and Papers Relating to the First Dutch War 1652*-1654, ed. S R Gardiner and C T Atkinson, six volumes, Navy Records Society, London, 1898-1930, II, 166, 234, 344

73 Wessex Archaeology Limited, *HMS London, Southend, Thames Estuary, Designated Site Assessment: Archaeological Report* (Ref.53111.03xxx, *Salisbury*, March 2011), 10, citing Port of London Authority, *London Gateway Dredging Plan – Wreck and Obstruction Categorisation, Wreck ID 343/2 W A 5029.* (unpublished, 2005).

74 Charles Trollope, personal communication and copies of private correspondence.

75 Wessex Archaeology Limited, *HMS London, Southend, Thames Estuary, Designated Site Assessment: Archaeological Report* (Ref. 53111.03vvv, *Salisbury*, March 2010), summarizes these developments.

76 E.g., *Diver Magazine*, June 2011, 17, in which the editors state unequivocally that the Dutch guns came from the *London*; and Wessex Ltd, *Archaeological Report*, 2010, which not only describes the guns as having come the *London* (Appendix III, 18), but reproduces photographs of them (Plates 1 and 2).

77 Nico Brinck, report to the Dutch Governmental Archaeological Department, Amersfoort, 2008, unpublished. Many thanks to Captain Brinck for this information.

78 TNA, WO 55/330, entry book; and WO55/1694, surveys of guns and stores, 31 December 1662.

79 W J van Hoboken, *Witte de With in Brasilie, 1648-1649*, Amsterdam, 1955, with English-language summary, 306 and 309; J E Elias, *De Vlootbouw in Nederland in de Eerste Halft der 17E Eeuw 1596-1655*, Amsterdam, 1933, 84 and n. 5. Dutch researcher Carl Stapel uncovered the more probable *Villeroy* identification (unpublished) in the Nationaal Archief in The Hague.

80 These data were compiled by comparing warships captured 1652-1654 (mostly extracted from numerous entries in Navy Records Society, *First Dutch War*) with James C Bender's vast unpublished *A Researcher's Friend: An Archival Source Guide for Dutch Warships in the 17th and 18thCenturies*, which he graciously made available. The relevant fleet ordnance lists cited by Mr Bender are all in the Nationaal Archief in The Hague: Staten Generaal 1.01.04 Lias Admiraliteiten Inv. Nrs 5550, 5551, and 5554 (the first two from 1652, the last from April 1653). Collectie Johan de Witt 3.01.17 Zeezaken 2774h, November 1652. Staten Generaal 1.01.06 Secrete Loketkas Inv. Nr 12561-125, 26 April 1653. All four of the ships listed in the text appear in two or more of these lists (the *Wapen van Rotterdam* appears in one). Ordnance data are lacking for some captured ships, but all except the *Gewapende Ruijter* seem too small to have had 24-pounders.

81 Bender, unpublished data list, gives armament and fates of Amsterdam directors' ships.

82 TNA, SP29/135, fo. 71 (see note 61). Mr Trollope's sources (personal communication) included Ordnance Office debenture books in TNA, WO 49/17-20, 37, and 42; an ordnance survey of 1609 in TNA, WO 55/1675; and an ordnance reassignment book of 1622 in TNA, SP14/133. For information on some of the *London*'s Tudor period guns, see C Trollope, 'The Guns of the Queen's Ships during the Armada Campaign, 1588', *Journal of the Ordnance Society*, Vol. 6, 1994, 23-38.

83 TNA, ordnance survey of c.1699, WO 55/1736, fo. 209v, survey no. 12636.

84 *Ibid.*, fo. 210, survey no. 12664.

85 See note 77 for guns 39, 40, and 41.

Acknowledgements

Although only one author is shown below the title, this article would have been impossible without a great body of research contributed by several scholars. Ordnance historian Charles Trollope generously supplied an extensive listing of sources relating to the *London* in the *Calendar of State Papers* and in the War Office volumes in The National Archives. I identified others, and instituted a search through some of the less explored War Office volumes. This was carried out at Kew (3,000 miles from my base) with brilliant success by seventeenth century researcher Richard Endsor, who painstakingly located and photographed hundreds of original documents from my own and Mr Trollope's reference lists. Dr J D Davies took the time to hunt down other sources, and Dr Peter Le Fevre offered manuscript papers which he discovered and correctly recognized as vastly more important than their corresponding *Calendar* summaries indicated. Graham Scott, lead archaeologist on the *London* for Wessex Archaeology Limited, furnished his team's intriguing progress reports on the wreck site. James C Bender very kindly allowed me full use of his highly valuable unpublished data and source lists for Dutch warships. Finally, Dutch ordnance expert Nico Brinck graciously shared his unpublished report detailing his examination of three controversial guns.

Frank L Fox, *of Birmingham, Alabama, is an independent researcher in naval history with particular interest in the Commonwealth, Protectorate, and Restoration periods.*

THE WOMEN OF RESTORATION DEPTFORD*

Richard Endsor

Abstract

To demonstrate the various influences women brought to bear upon the Deptford yard in the mid-seventeenth century, the cases of eight ladies of differing status are outlined. A comment is made about chips allowance. It is concluded that while a social division certainly existed, it seems to have been less rigid than that in ensuing centuries.

Over a period of several years researching the history of the warship *Lenox* launched at Deptford in 1678, I came across many interesting characters, including a number of women, whose stories were intimately linked to the town and dockyard. They did not of course live in isolation but with their families and friends. Dockyards are normally regarded as being the province of the male gender, but there is evidence that from time to time the distaff side did have an influence on yard activities. This is not altogether surprising in itself, but what is unexpected is the variety of ways that women did have a role to play.

Deptford dockyard was founded during the reign of Henry VIII in 1513 on the narrow Thames near London, but by the Restoration period was primarily a construction and repair yard. It was situated about a half mile upstream from Greenwich where the Queen's House stands. Nearer the river the King's House, now part of the Royal Naval College, was being built to replace the old Royal Palace of Placentia. Deptford was a small town still surrounded by green fields with the dockyard covering the Thames waterfront. The landward side was bounded by the 14th century church of St Nicholas and Sayes Court, the gentlemanly home of John Evelyn. The grounds of Evelyn's home covered 100 acres and were laid out with gardens, walks, groves, enclosures and plantations.[1] In between lay the tightly packed houses of the community most of whose lives were devoted to the yard. Its facilities included a great dry double dock and two building slips or 'launches', known as the Great Launch and the Lesser Launch. It also contained a great storehouse which held stores for the whole Navy.

Louise de Kérouaille

The first example of the importance of women in Restoration Deptford is demonstrated by King Charles and his love for Louise de Kérouaille, Duchess of Portsmouth. On 12 April 1678, with much of the court in attendance, Charles came down the River Thames in barges from the Palace of Whitehall to attend the launch of the third

Fig. 1. Louise de Kérouaille (1649-1734). A late seventeenth or early eighteenth century painting on board. Author's collection.

---

* *Paper presented at the fourteenth annual conference of the Naval Dockyards Society held at the National Maritime Museum, Greenwich, on 17 April 2010. Theme: Pepys and Chips. Dockyards, Naval Administration and Warfare in the Seventeenth Century.*

rate ship *Lenox*. It is almost certain that Louise, the King's mistress and her five-year-old son by the King who they named Charles Lenox, was amongst them. Louise lived in 'glorious apartments at Whitehall'.[2] However, as a French Catholic Louise was unpopular among the general populace at a time when fears of popish plots were rife.

Louise met King Charles years before when she first came to England as maid-of-honour to his sister, Princess Henrietta, who had married Philippe d'Orléans, the younger brother of Louis XIV. She became one of Charles's mistresses, but in 1674 contracted a venereal disease from him, suffered a miscarriage and became very ill for a considerable period.[3] Needless to say she was extremely upset and did not see Charles for over two years while he embarked on a number of other relationships. By 1677 Charles and Louise had both recovered their health and had recently reconciled.[4] Naming the first of the new ships after their son was a gesture of Charles's love for Louise and she remained his primary love until the end of his life.

The Shish Family

Coming down the social order at Deptford we find the Shish family of shipwrights. They had been carpenters at Deptford for over 300 years.[5] Jonas, the patriarch of the family was born in 1605 and had three sons, two of them, John and Thomas, became eminent Master Shipbuilders for the King while the third son, Jonas Junior, built ships at the family's private yard. The women of the Shish family are typical rather than exceptional, but gave and received the love of their men. Little is known of Jonas's early years, but he married Elizabeth Francis in Saint Gregory's near Saint Pauls, London, in 1636 when he was 31 and she 17 years old. During their long marriage they had eight children of whom three sons and three daughters survived them.[6] He was Assistant Master Shipwright at Deptford and Woolwich at the Restoration in 1660,[7] establishing a good reputation as a builder of successful ships.

John Evelyn thought highly of the family and thought Jonas, apart from his honesty, remarkable for bringing up his children well. Jonas and Elizabeth's eldest son, John, was born in 1643 and was educated in shipbuilding. In 1668, when he was 25 years old he became his father's Assistant following the promotion of Jonas to Master Shipwright at Deptford.[8] John married Mary Lake, his 'dear and loving wife' on 22 November the same year.[9] A little over ten months after their wedding their first son was christened at St Nicholas's church, Deptford. Unfortunately the child, named Jonas after his grandfather, seemed to have died in infancy.

By 1671 he and his wife had another son named Kendrick. Two years later at the age of 32, John exchanged positions with his

Fig. 2. The launch of the *Lenox* at Deptford in 1678. Painting by author.

father and became Master Shipwright at Deptford, while his 27-year-old younger brother, Thomas, became his assistant. Thomas's wife Rebecca had given birth to one child and would have at least seven more.[10] They had a lucky escape during the evening of Sunday 16 January 1681 when Thomas was in his chamber while his wife lying in her childbed. He noticed an unusual amount of smoke appearing through the jams of the chimney and while putting out the fire and breaking up the hearth found the joists below to be on fire which he immediately quenched.[11]

When John Shish launched the 90 gun *Neptune* in 1683, John Evelyn, who lived next door described him as 'my kind neighbour young Mr Shish',[12] and in 1685 mentioned 'the great respect I have to Mr Shish'.[13] John Shish showed typical concern for the women of his family, when one of his sisters was dangerously sick in Kent and she requested he come and visit her, he immediately wrote to the Navy Board desiring leave to be gone from Deptford as soon as possible.[14] In 1679, in an act of public duty, he and some of the dockyard officers became Feoffees, a kind of trustee, to the New School House, East Greenwich.[15]

Elizabeth Shish, the wife of Old Jonas died on 5 March 1678, an event that must have deeply saddened him. She was 59 years old and had twelve grandchildren. Jonas himself lived on until 7 May 1680, dying at the age of 75. During his last sickness he wrote his own gloomy puritan epitaph which can still be read: 'Once I was strong but am intombed now to be dissolved to dust and so must you'. Old Jonas was buried at St Nicholas Church Deptford, in the coffin which he had lying by him for many years. It was his custom to rise in the night and pray kneeling in it. He had gained the respect of many people and at the funeral John Evelyn and three knights acted as pall bearers.[16] Jonas was spared the tragedy that befell his sons. Thomas died in 1685 aged 37, the same year Kendrick, the only surviving child of John and his wife Mary, who died at the age of 14.

John died at the age of 43 in October 1686 leaving his wife Mary, his 'dear and loving wife' as sole executrix and he left her a considerable fortune. There were tenements and premises in Canterbury, tenements, gardens, closes, lands and premises in Erith, tenements, yards and premises in Deptford, freehold lands, closes, meadows, pastures, pasture woods, underwoods, other various premises and a number of acres lying in Hatfield Broad Oak. Mary was also to have all his goods, chattels, and household stuff including his plate, rings and jewels. Apart from all this there was Flagon Rowe, a street in Deptford, which he had bought jointly with his sister in law Rebecca Shish and he now left his share to her. He left smaller amounts to other members of his family and to the poor of Deptford. Perhaps the most poignant request in his will was for his wife to look after the education and upbringing of the infant son of his dead brother Thomas, and take as much care of the boy as if he was her own. He also left him £300 put in trust until he was 21 years old. The infant, named Thomas after his father, was presumably an orphan.[17]

The only member of the family who lived to a good age was Old Jonas, but longevity at Deptford was not confined to him. If women survived the dangers of childbirth, plagues and epidemics they could far outlive their men folk. Maudlin Auger from Deptford was buried in December 1672 aged 106, while Catherine Perry was buried in December 1676 having lived to be 110. She would have remembered, as a 22-year-old young woman, the ships being fitted out at Deptford to oppose the Spanish Armada.[18]

Mrs Bagwell

The puritan Shish family was very different from the flamboyant Samuel Pepys. They were also very different from the Bagwell family of shipwrights who also lived at Deptford. Owen Bagwell was the Foreman and his son, William, a shipwright and warrant officer Carpenter of the *Dolphin*, a small vessel laid up there. His interesting wife has been made famous by Pepys's diary in which he always calls her Mrs. Bagwell leaving her first name a mystery. However, in 1684 there was an Anne Bagwell, wife of William, a Deptford shipwright who used the alias Bayley. She was mentioned in a dispute over a mariner's pay ticket belonging to Abraham Moody of the *Woolwich*.[19] This may be the Mrs Bagwell of Pepys's diary.

Pepys records in his diary that on 9 July 1663 while visiting Deptford yard on Navy business he sought out William Bagwell who had a pretty wife. Pepys walked with him alone to get to know him and found an excuse for his wife to visit him at his office. A couple of weeks later the Bagwells invited Pepys into their house to drink some wine where he found Mrs Bagwell to be virtuous and modest. After Pepys's visit to their home and up until the end of 1663, Pepys had little contact with them other than occasionally being asked if he could find a better ship for William.

During their first year of friendship, Samuel Pepys and Mrs. Bagwell met infrequently and in all that time the most he did was stroke her under the chin during one of her visits to the Navy Office on her usual mission in seeking promotion for her husband. Pepys seems genuinely to have thought her modest and therefore made no advances. As he got to know her better things took a turn for the better, or worse depending how you look at the situation. On 3 October 1664, he kissed her and although she rebuked him, he thought she liked it.[20] Thus encouraged, Pepys gradually increased the earnestness of his advances and the frequency of seeing her, sometimes arranging to meet at alehouses to eat and drink. For sixteen months after they first met the most he did was caress her. Then, on 15 November they met at Moorfields Alehouse and after eating and drinking, by degrees he arrived at what he would, with great pleasure, even though she protested and was troubled at what he did.

Once this line had been crossed the relationship developed into the most immoral he would have with any of his many mistresses. Most women were clearly having fun with an attractive man of sharp wit who relished pleasure. However, although Mrs. Bagwell may well have enjoyed the relationship she had other motives, the advancement of her husband. Pepys no longer felt the need to meet and woo her in alehouses but often made arrangements to visit her home after dark.

During the following winter he met her a number of times. Then, on the 20 February 1665 he wrote to the Earl of Sandwich requesting a better ship for Bagwell. He then went straight away by boat to Deptford with the news but waited till dark before going to her house. He sought a reward for his efforts and after a great deal of difficulty eventually had his way, then after eating and drinking he went home. Such behaviour had its price for the next day he wrote of having a mighty pain in the forefinger of his left hand from a strain it received in struggling with her the night before. After this experience he didn't see her again until 19 July and although he stayed until midnight he didn't get anywhere with her. However, things got back to normal when she visited him at his office on 8 August and he did what his heart longed to do with her.

The depth of Samuel and the Bagwell family's decadence was revealed during the summer when Pepys met her father-in-law, the foreman shipwright, Owen Bagwell. He encouraged Pepys to visit her, which he did and proceeded to have his way with her.[21] Shortly afterward Pepys met both Mrs Bagwell and her mother-in-law at the Deptford yard back gate and then went alone with his lover to her house and did whatever he wanted with her.[22]

Their affair continued throughout 1665 during the great plague and even though it was each side of Bagwell's house he still visited her.[23] During June 1666 he even had the courage to visit her and do what he wanted even though her servant had died of it.[24] Her husband returned from sea in the *Providence* a week later but this made no difference to the frequency Pepys saw her.[25] On 12 September 1666 he visited her after dark and went to bed naked with her intending to stay all night. But after he had what he wanted he felt revulsion and being uncertain of her husband's return, left for home. The excitement of the affair was clearly starting to wane with familiarity.[26] The affair continued into 1667 when Pepys arranged for William Bagwell to go to Harwich to become the Carpenter of the new third rate ship *Rupert*.[27] The frequency of

Pepys seeing her then fell away for the rest of 1667 until the end of the Diary period on 31 May 1669. In all Pepys records seeing her about 37 times with two periods of high frequency, the first from October 1664 till March 1665 and again during June and July 1666. Shortly after he finished his diary his wife, Elizabeth Pepys, died, although Pepys's career developed and he was appointed Secretary to the Admiralty in 1673.

Pepys continued his friendship with the Bagwells. In a testimonial he wrote in October 1677 concerning the advancement of Bagwell, then Carpenter of the *Resolution*:
do assure you that as well from the character you are pleased to give him as my own many years knowledge of him. I both have and shall endeavour to do him a good office'...'Mr Bagwell hath had the opportunity of having his name and character by this occasion presented to his Majesty and will stand very fair for the first good turn that shall offer for him.[28]

Sure enough Pepys secured a warrant for the appointment of Bagwell to the position of Master Carpenter of the first rate *Prince*.[29] Then, as a Master Carpenter with experience of shipbuilding at Deptford, Bagwell was sent to Bristol to supervise the building by contract of a third rate of the 1677 programme built by Francis Bayley. While at Bristol he wrote a letter saying his wife was keeping him informed of the great endeavours the Navy Board was making on his behalf.[30] At the same time William Bagwell wrote another letter requesting the advancement of his brother, Francis, from the position of Carpenter of the sixth rate *Greyhound* to Master Carpenter of the ship he was supervising being built.[31] By some means, of which we can only guess, Mrs. Bagwell was still in a position of some influence. In April 1681 Francis Bagwell confidently wrote to the Navy Board himself requesting his servant be entered and borne on the Deptford yard paybooks.[32] He wrote again in September requesting their Honourables to grant him a happy favour, having a great charge of small children, and to allow his servant to remain at the yard until his ship be ordered to sea.[33]

Ten years later on 7 January 1687, 25 years after they first met, the 54 year old Pepys wrote to William Bagwell a letter saying that he was his friend and will continue to be so. He then advised that his wife should not waste time in visiting him as he knew her request and that she will hear when the favour is done. The favour was, of course, the same one he had heard during all their years of friendship. Explaining this to William Bagwell he went on
This I thought fit (out of my old friendship to you which I have no reason to alter) to say to you for removing the apprehensiveness you seem to be under of my backwardness on occasion of the late vacancies of a Master Shipwright's and Assistance places.[34]

There is no way of knowing what sort of relationship he then had with Mrs Bagwell. At 54 he was probably still capable but whether he still found her attractive when she was roughly 45 years old we may never know. In spite of the unsavoury nature of the relationship between Pepys and the Bagwells it withstood the test of time. William Bagwell went on to become Assistant Master Shipwright at Chatham in 1689, Master Shipwright at the small yard at Sheerness in 1694 and finally, Master Shipwright at Portsmouth a year later where he stayed until his death in 1698.[35] It is no real surprise to find women like Mrs. Bagwell making their way in the world the best way she could. The backlash led by King Charles after the puritan regime and a sense of live for today as there may not be a tomorrow, especially during the plague, must certainly have played their part.

Such goings on were not particularly unusual in the seventeenth century and Mrs Bagwell was not the only woman in Deptford who enjoyed a frolic. A ballad relating to another woman from the town during the same period survives. Her husband appears to have been a watchman who probably worked for the Dockyard.[36]

The Deptford Frollick; or a Hue and Cry after the Shag-Breeches

Young women all, both great and small,
That handleth Pot and Pail,
Forsome I hear, and greatly fear,

Do oft play with their tayl.

*One night when blustering winds blew
And busy was the sky
Tho' I was feeble weak and old,
A watching then went I:
But cruel fate did prove unkind,
My grief did then begin;
And quite contrary to my mind,
At anger he got in.*

*Two keys unto my house I had,
As I did think it fit,
But now it makes me almost mad
I had so little wit;
For when a watching I was gone;
A wanton licentious man
Unknown to me got in.*

*Into the bed straightway he went,
And hugg'd my loving Wife,
Who used to give me hearts content,
I loved her as my life,
And grieve to think she should commit
So foul and grosse a sin,
And let him do what was not fit,
When she had let him in.*

*As they in sweet embraces lay
I chanced to return,
And spoiled the game which they did play,
For which my wife did inform
She told me she was wonderous ill,
And thus she did begin
With shrieks & groans she made her moans
Cause she let him in.*

*I willing was to go to bed
And off my breeches threw
She told me she was almost dead,
And knew not what to do:
Dear love (quoth she) a cordial get,
My pains, my pains afresh began:
I little thought she was so naught
To let another in.*

*Away went I most willingly
For my dear spouses sake
A pair of breeches on put I
Which proved a mistake;
I to the Apothecaries went
Thinking her love to win.*

*A cordial made I asked to have,
Not thinking who got in.*

*A cordial was prepared for me
Then thus I did reply:
At present I cannot pay thee
But yet assuredly
Tomorrow I will come and pay:
My pocket I felt in,
And there behold was store of Gold
The youngster had brought in.
The Apothecary he did know
The breeches I had on,
And then he said full well he knew
The things I stared upon
How I by these Shag-Breeches came:
to pause then did begin
At last thought I assuredly
She let some Gallant in.*

*Away went I most furiously
This trick to think upon;
But when I came with grief and shame
The youngster he was gone
I had his watch and money to
And I the things did win
But I am mad and monstrous sad
That she had let him in.
Take warning, all both great and small
In women ne'r confide
For some pretend to their lives end
They constant will abide
Past all relief unto my grief,
I know they are prone to sin
And when you'r gone some other men
Sometimes may happen in.*

Susana Beckford

Another woman who had intimate relationships with the officers of Deptford yard was Mrs Susana Beckford. Her relationship was of a totally different kind from Mrs. Bagwell's, as she was the legitimate and approved supplier of ships' ironwork to Deptford and Woolwich yards. She had been married at least once before becoming Mrs Beckford and had two sons by previous marriage.[37] Her husband, Thomas Beckford, a locksmith and ironmonger, had supplied ironwork to the yard since at least 1674,[38] but fell dangerously ill on 14 December 1675. As he lay close to death, with his wife crying and grieving and fearing for her future he told her 'my dear, do not distress yourself for I give all that I have in the world to you'.[39] He died hours later and news of his death spread quickly throughout Deptford and the Navy administration. The next day John Shish wrote to the Navy Board in London on her behalf:

The Widow Beckford returns you many thanks for all your kindness and would desire you that her two sons may have leave granted them to come up from Chatham to Deptford to the funeral of their father which will be a Friday next. The one is Carpenter of the Antelope, namely Henry Hallbert and ye other is John Hallbert Carpenter of the Revenge and would desire leave a week's time for Henry to pass and make up some accounts and three days for John.[40]

Luckily for Susan, as she was usually known, Thomas Beckford's verbal will had been witnessed by two other women, Deborah White and Anne Clapp. The verbal will was legally accepted and Susan Beckford duly inherited the business, and recovering from the loss of her husband took the opportunity to run it herself.

On 3 September 1677, John Shish, the Master Shipwright had dealings with her. At the time he was responsible for hundreds of workmen building two third rate ships. He had few problems handling the workmen but dealing with Susan Beckford was more than he could handle; he wrote to the Navy Board:

There is one brass plate lock belonging to His Majesty's Ship the Greyhound which wants a new key. The lock is very good, but Mrs Beckford refuses to make a new key to the said lock. The reason, as she informs me, is that she hath not a price answerable for such a key which I humbly leave to your Honourables consideration.[41]

Mrs Beckford was of course fully aware of the commercial advantage of supplying a new brass plate lock rather than just a key. Interestingly, John Shish allowed £5 10s for several locks and hinges in his total estimate of £176 10s for repairs to the *Greyhound*,[42] and is the same ship that Francis Bagwell sought promotion from through the efforts of his sister-in-law, Mrs. Bagwell.

During 1677 Susan Beckford was under the impression she was not being paid enough for her ironwork and became a thorn in the flesh of other officials at Deptford. Frank Hosier, the Clerk of Control, wrote to the Navy Board complaining she demanded a copy of her rates of payment, but he did not give her one and wrote:

If your Honourables think fit to let her have one I may be excused, 'twas her extreme importunity has made me give you this trouble for which I humbly beg your Honourables pardon.[43]

A few days later, on 22 June 1677, she wrote to the Navy Board herself complaining

I received my bills and find I am not allowed the prices for my commodities as is according to contract, and also there is many things, that I cannot afford so cheap as they are rated. I beg Your Honourables perusal of the enclosed and crave leave to remain, Your Honourables Most Humble Servant. Susann Beckford.

Her letter is in the same educated and well written hand as the signature, indicating she had been well educated. The Navy Board received her letter the same day and acting in all fairness delivered her bills to Mr Barbour to examine; he found them less than another supplier, called Burton. Other not comparable bills were sent to the Master Shipwright, John Shish for his opinion.[44] Not happy, she somehow managed to persuade the Deptford officers to issue bills that did not agree to her contract and receive money for them. This was soon noticed by the Navy Board Surveyors office which wrote to William Fownes, the Clerk of the Cheque, and other officers at Deptford, saying

that ye enclosed bills of Mrs. Beckford should contain any goods not agreeable to contract it is required that Mr Shish and ye Officers concerned do value those particulars unfilled... and never to receive any goods of hers for ye future, or of any person else, unless agreeable to contract.[45]

Although clearly a handful to deal with, she delivered over £461 worth of work for the new ships of 1677. This in today's terms is nearly half a million pounds. She also supplied goods for the repair of old ships and by April 1680 her outstanding bills came to £800. With such a huge amount of money owed she wrote to the Navy Board reminding them of the value of goods she had delivered to Deptford and Woolwich yards for repairs, and that she was daily supplying goods for some ships fitting out for sea at Deptford. She requested that some of the debt be paid in order for her to continue supplying the ships.[46] Over the years she continued supplying ironwork such as curtain rods, iron knees and rudder irons for longboats and gunport rings until at least 30 June 1685, supplying goods worth £77 13s 5p for the preceding 3 month period.[47]

If the Deptford Dockyard officers had a difficult time at the hands of Susan Beckford they did not suffer as much as their next-door neighbour, John Evelyn. On 3 August 1685, Evelyn, from his house at Deptford, wrote a letter full of sadness to his friend Samuel Pepys. His daughter Elizabeth had eloped with the nephew of Sir John Tippets, to the amusement and knowledge of the whole of Deptford. Now she was dying from smallpox and Evelyn's wife had lain ill for two days with a fever caused by her sadness. In his letter Evelyn blamed Susan Beckford as an assistant in the connivance of his daughter's elopement.[48]

It appears tragedy struck Susan Beckford's family the same year when Christopher Beckford, also of Deptford, died aboard the *Phoenix*. He was probably her son or brother-in-law.[49] Perhaps finally disillusioned by her loss, she petitioned King James II in December 1685 to have her business joined by warrant of approval with that of Isaac Loader, the Anchorsmith of Deptford. The Navy Board agreed there would be no inconvenience and recommended her request be granted.[50] A warrant was duly made out dated 26 March 1686 signed by James II and Samuel Pepys.[51]

We may think she was remarkable in her boldness and strength of character in dealing with the men of her age. She probably was in many ways, but her forthright attitude is also reminiscent of another character we know very well, Samuel Pepys. Their confident, brash and spontaneous actions are features of so many men and women of the Restoration period.

Anne Pearson

Another widow who worked in Deptford dockyard in the 1670s was Mrs Ann Pearson. With care and pains she laid the deadly poison for destroying the rats in his Majesty's stores at Deptford and Woolwich - not an attractive occupation so soon after the devastation caused by the plague a few years before. She petitioned the Navy because her certificates confirming her work for a previous six months had been mislaid or lost and she had not been paid.[52] Shortly after her petition, a bill was made out to her for £14,[53] meaning her yearly earnings were £28, a good sum and about the same amount earned by a skilled shipwright. In her humble petition, she calls herself a poor widow in very great want.

Ann Mavely

Although her links are with Chatham rather than Deptford, the letter of a pregnant woman, Ann Mavely, is worth the telling. Her husband had been recently pressed to work in the dockyard at Chatham causing her to write to the Navy Board during June 1677:

Please for to order that my husband John Maverly, a caulker, may have a month's liberty before he goes into His Majesty's yard at Chatham being imprest last Wednesday in respect I am very big with child and look every day and having several things to do which I am not able to perform without my said husbands assistance. I humbly beg your Honourables compliance therein.

The word 'Mavely' is written in the letter and is the same as the signature indicating she had written the letter in her own very neat hand. After consideration the tolerant Navy Board granted John Mavely one week's paternity leave.[54]

The Widows Gutteridge and Gurling

Thomas Lewsley was the Purveyor of Timber based at Deptford who also worked on behalf of Chatham and Woolwich yards. He was extremely knowledgeable in his trade and Pepys had consulted him about the quality of supplies since at least 1664.[55] In 1677 he received instructions from the Navy Board to go into Suffolk to view plank and negotiate prices and terms. Lewsley rode through the leafy countryside to Woodbridge where he met three timber merchants. Two of the merchants, the widows Gutteridge and Gurling, proved tough ladies to do business with. They had generally good wood, reported Lewsley, but he could not persuade them to sell it on any other terms than ready money.[56]

Chips

Since this conference is themed Pepys and Chips and we have examined Pepys's relations with women of Deptford, we ought also to look at women and chips. One of the benefits workmen and their families enjoyed was chips – the pieces of timber not suitable or large enough for shipbuilding and unsuitable as fuel to boil pitch. Men were allowed to take them home but the seemingly harmless allowance caused endless problems. Shipwrights might cut up large timber to be useless except as a chip. It had long been a problem; in 1664, one sawyer at Deptford had his wife bring him a breakfast two or three times a morning and leave carrying out chips every time. The enterprising couple then sold off the timber for between £2 and £5 each month.[57]

Sir John Tippets, the Surveyor of the Navy, discharged Richard Lawrence, a labourer, for splitting good timber into chips. Unfortunately for him he had worked in the yard for a year and a quarter and was owed most of his wages, which would now be forfeit. He appealed to the Navy Board for tolerance saying that he had a wife and child and his creditors would have him thrown in prison. Influenced by the plight of his family they generously ordered he should be included amongst the men who were generally discharged and allowed to receive his wages.[58]

It was also the custom at Deptford to allow the poor people of the town to enter the yard on Wednesday and Saturday to gather small chips. Over time the scheme gradually fell into abuse as the wives, children and friends of the workmen joined the poor people carrying out good timber along with chips. The situation deteriorated so that eventually the workmen, ever willing to take maximum advantage, cut up useful timber and hid it for collection later. In 1698 a search revealed many of the abuses including lengths of good timber hidden in holes cut into the earthen sawpits. For the future, the Navy Board ordered that no wives, children or friends of workmen were to be let in the yard and the sawpits were ordered to be brick lined.[59] Another way the poor people of Deptford were able to earn money was by unpicking old Navy rope or 'junk' into oakum for use as caulking. They were paid directly by Thomas Turner the Storekeeper who was later reimbursed by the Navy Board.[60]

Conclusion

What is so surprising about the woman of Restoration Deptford and indeed men and women from widely differing social class, is the interaction between them. It is known that Pepys regarded members of his household as junior members of his family. He may molest women, who were willing, and perhaps chase them round the house if they broke something, but he certainly knew who they were, whether they could sing, how honest they were and generally took an interest in them. He probably was not exceptional in his attitude being a man of his times, albeit one with exceptional talent who lived life to the full. How different this was during the later eighteenth and nineteenth centuries when there was a much greater gulf between the classes. This attitude is reflected by the discipline in the Navy where a much greater tolerance is evident in the seventeenth century. Whatever we think of the way Mrs. Bagwell or Susanna Beckford behaved, they were arguably in control of their own destinies and it is only in comparatively recent times that women have again been able to achieve this.

References

1. G Bell, *Diary of John Evelyn*, 1902, 17 January 1652.
2. *Ibid.*, 24 January 1682.
3. Bryan Bevan, *Charles the Second's French Mistress*, 1972, 63.
4. Rosemary Baird, *Mistress of the House*, 2003, 65-73.
5. Bell, *op. cit.*, 3 March 1668.
6. The Shish Memorial, St Nicholas Church, Deptford.
7. Robert Latham and William Matthews, *Diary of Samuel Pepys*, Vol 10, 397.
8. C Knight, 'Master Shipwrights', *The Mariner's Mirror*, 81, 1995, 412.

9 www.familysearch.com

10 *Ibid.*

11 TNA, ADM 106/358 f269.

12 Bell, *op.*cit., 17 April 1683.

13 Guy De La Bédoyère, *Particular Friends. The Correspondence of Samuel Pepys and John Evelyn*, Boydell, Woodbridge, 2005, 154 (3 August 1685).

14 TNA, ADM 106/370 f348.

15 Nathan Dews, *The History of Deptford*, 1884, 169.

16 Bell, *op.cit*., 13 May 1680.

17 TNA, PROB 11/385.

18 Dews, *op.cit.*, 316.

19 TNA, ADM 106/3540.

20 Latham and Matthews, *op. cit.*, 3 October 1664.

21 *Ibid.*, 12 August 1665.

22 *Ibid.*, 22 August 1665.

23 *Ibid.*, 5 October 1665.

24 *Ibid*, 13 June 1666.

25 *Ibid.*, 23 June 1666.

26 *Ibid.*, 12 September 1666.

27 *Ibid.*, 4 March 1667.

28 TNA, ADM 106/37.

29 TNA, ADM 106/38, 9 March 1678.

30 TNA, ADM 106/321 f21.

31 TNA, ADM 106/341 f132.

32 TNA, ADM106/353 f138, William Bagwell to Navy Board.

33 TNA, ADM106/353 f139.

34 Pepysian MSS, Adm Letters XI, 531-2. Reproduced by Arthur Bryant, *Saviour of the Navy*, 166-7.

35 Knight, *op.cit.*, 411.

36 Author's collection.

37 TNA, ADM 106/310 f352.

38 TNA, ADM 20/22.

39 TNA, PROB/11/350.

40 TNA, ADM 106/310 f352.

41 TNA, ADM 106/323 f323.

42 TNA, ADM 106/3118, p275.

43 TNA, ADM 106/324 f368.

44 TNA, ADM 106/321 f492.

45 TNA, ADM 91/1 23, July 1677.

46 TNA, ADM 106/347 f118.

47 Admiralty Library, Portsmouth, MSS 10.

48 Guy De La Bédoyère, *op.cit.*, 154.

49 TNA, PROB 4/21452, 11 December 1682.

50 TNA, ADM 1/3555, p.531.

51 TNA, ADM 106/46.

52 TNA, ADM 106/3539.

53 TNA, ADM 18/60 p.267.

54 TNA, ADM 106/326 f154.

55 William Matthews and Charles Knighton, *Samuel Pepys and the Second Dutch War*, Navy Records Society, 133, 1995, 105.

56 TNA, ADM 106/326 f40.

57 Matthews and Knighton, *op.cit.*, 69.

58 TNA, ADM 106/3538.

59 TNA, ADM 49/132, 7 October 1698.

60 TNA, ADM 18/61, 21 January 1678.

Richard Endsor *is an artist with a particular interest in the Restoration period.*

INTERSECTIONS OF INTEREST; A PROSOPOGRAPHICAL ANALYSIS OF RESTORATION PRIVATEERING ENTERPRISE*

Richard M Brabander

Abstract

Privateering, not to be confused with piracy, was a regulated business enterprise which contributed to state coffers as well as the pockets of private investors; it nevertheless did involve maritime predation. The activity assisted the emergence of the fiscal-military state which saw the necessity for regulating private behaviour to support standing armies and navies. By examining individual warrants, affidavits, captains' declarations and bonds for each privateer ship – the prosopographical approach – the paper analyses the reciprocity between privateering enterprise and, firstly, the Company of Royal Adventurers of England Trading to Africa, and secondly, the Royal Navy. Highlighting the nature of privateering, and indeed of Restoration society, Pepys' somewhat ambiguous role as an investor and a public servant is considered.

Introduction

The Privateers in our Wars are like the Mathematici in old Rome. A sort of People that will always be found fault with, but still made Use of. [1]

Privateering, throughout history, has suffered from an image problem stemming from the ubiquitous myth that privateering was synonymous with piracy, or more benignly that privateers were only one step removed from banditry. Much attention has been given to the popularized accounts of lawless maritime predation in the Americas. Little attention has been given to privateering enterprise as it actually existed in Europe: a respectable and regulated business enterprise with centuries of legal precedent that served the needs of both private and public spheres in the early modern period. Equally to blame for a historical misrepresentation of privateering is the persistence of Alfred Thayer Mahan's nineteenth century interpretation of naval history. Mahan trivialized privateering with his influential thesis that sea power (and hence world power status) is the result of a state maintaining a large battlefleet that exists to seek decisive engagements, eschewing commerce warfare.[2] This present study will not delve into these theoretical and nomenclature debates but will briefly synthesize recent scholarship to create a framework for a more modern, accurate understanding of the nature and established practices of privateering as it existed in Europe (in this particular case, England) during the Restoration period and beyond.

Briefly put, Restoration privateering is significantly far removed from the 'privateering' of the Elizabethan age for several critically important reasons. To begin with, the word 'privateer' does not enter the English language until roughly the middle of the seventeenth century, and the nature and conduct of men such as Sir Francis Drake and the 'Sea Dogs' of the Elizabethan age hark back to a medieval practice (starting in the late thirteenth century) known as the doctrine of reprisal or *lettre de merk*; which by its strict definition applies only to a specific aggrieved individual who acquires authorization from his state to obtain a private 'redress' for his losses on the high seas due to the unlawful actions of a foreign subject.[3] Since the state itself is not at war, it may not authorize a private ship to conduct public war against a rival state in its name; hence a *lettre de merk* for 'corsairs' (perhaps the most accurate term for men such as Drake) entails a private individual war that could only be issued during a time of peace.[4] The aggrieved claimant had to offer proof of

---

*Paper presented at the fourteenth annual conference of the Naval Dockyards Society held at the National Maritime Museum, Greenwich, on 17 April 2010. Theme: Pepys and Chips, Dockyards, Naval Administration and Warfare in the Seventeenth Century.

both the losses incurred and the unlawful nature of the seizure. However, it was not until 1586, when the High Court of Admiralty had the power and authority to formally condemn prizes (and adjudicate them as lawful), that any agreed upon international standards of legality for any such reprisal seizure on the open seas were met.[5] During the sixteenth century this use of 'private war' was the norm; European states did not have the wherewithal to control or regulate violence on the high seas as they lacked the bureaucratic apparatus and public backing to support the military expenditures necessary to sustain a large permanent state navy. Corsairs were thus in a real sense a private substitute for a public state navy. Historians have long noted the discernible shift from the significant use of private non-state actors/resources as a substitute for a public navy during the Elizabethan age to that of a predominantly national/public navy towards the end of the seventeenth century. Understanding the nature of this evolutionary shift as it relates to privatized maritime predation in the seventeenth century has traditionally been poorly understood (no doubt due to the lexicon confusion over the word *privateer* itself!). Nonetheless, these developments that did begin to occur, in the case of England particularly after the Interregnum, are critical requirements for a 'privateer' to exist in a true legal reality as a private non-state actor authorized by a state that enjoys a near monopoly on the application of violence. Thus, armed with new bureaucratic apparatus to enforce regulations and with the shift from the primarily private navy of the Elizabethan age to the predominantly public navy nearly complete by the Restoration period and beyond, the state now had the authority and wherewithal to properly regulate these private entrepreneurs who wished to engage in maritime predation in a privately owned ship in times of openly declared public war against a specified enemy's commerce, under the strict control and guidelines of the state as a licensed 'privateer'.[6]

It is therefore an anachronism to label sixteenth-century corsairs authorized by letters of reprisals as 'privateers'. Corsairs who sailed with an individual letter of reprisal throughout the Elizabethan age would flagrantly ignore the terms of their *lettre de merk* by seizing goods on the open seas far in excess of the reprisal claim; as long as England was in an undeclared war with Spain, these corsairs and their depredations served the interests of the state. However, once England made peace with Spain in 1604, English corsairs took little notice of the fact and continued their now deliberately piratical maritime predations. Therefore, these corsairs had become a major diplomatic liability, requiring King James I to issue several decrees outlawing all 'piracies' on the sea.[7] Individual letters of reprisal were rarely issued after this radical change in policy during the early Stuart era, and to encourage this new policy of restraining privatized maritime predation during times of peace, the aggrieved individual now had to show definitive proof of actual damages in order to obtain a valid letter of reprisal. By the time the Admiralty imposed a new vigorously enforced restriction on the doctrine of individual reprisal in 1649, which made it a crime to seize goods in excess of a reprisal claim, the *lettre de merk* had for all intents and purposes become an anachronism and had largely fallen out of use.[8]

Therefore, throughout the course of the Stuart age there is a marked transition from the era of the letter of reprisal corsairs to that of a new era of privatized maritime predation: 'privateering' which would emerge in its more modern form as not a *substitute* but a vital strategic and economic *supplement* to the already stretched and costly battlefleets of the period and become a well-established and important feature of England's strategic maritime interests through the nineteenth century.[9] With the outlines of modern privateering in place by the Restoration, the English state took every opportunity to sanction these opportunistic purpose-built private warships (non-state actors) to engage in maritime predation on the shipping of a public enemy during times of declared war. Privateers were typically dispatched as the first indication of an impending war or even as a tool to provoke war, as demonstrated by the sending out of sixteen privateers in February of 1664[5], a month before war was declared on the Dutch.[10]

It cannot be emphasized enough that these privateers of the Restoration period and beyond were far removed from their sixteenth-century corsair predecessors. The actions of the privateer would now follow a highly regulated system that dictated the actions of the privateer, from the signing (before the ship set sail) of the warrant and bond otherwise known as a 'letter of marque' (once again the nomenclature harks back to medieval terms) to the proper adjudication of each captured enemy ship by the High Court of Admiralty.[11] Of particular importance is the bond, which was the method by which the state ensured the proper conduct of the privateer; the consortium of individuals who owned the privateer ship had to come up with £4000 (an enormous sum for its day) to be held by the state for the duration of the voyage. If the privateer acted inappropriately, or turned pirate, the bond money would be forfeit to the state.[12] This regulatory tool of the state and the fact that the captain of the privateer ship was most often one of the consortium owners (another strong motivator for the privateer to act appropriately as the captain's own money was at stake) meant that privateering in European waters was indeed seen as respectable and far removed from the lawlessness of maritime predation in the Americas. It is this form of 'gentlemanly' privateering that is the subject of this article.

What makes privateering enterprise an important subject far beyond its maritime roots is its emblematic connections to larger processes in early modern history. This understudied topic has relevance not only for our understanding of Restoration England, but for larger issues in early modern European history such as analyzing the evolution of the 'fiscal-military state' and the complex relationship between private citizens and the burgeoning early modern state. Tracing the evolving relationship between the individual subjects and their state (whether monarchy or republic) is crucial to understanding the overall transformation of Europe from a relative backwater periphery at the start of the Early Modern period to its progressive domination of the Atlantic World and later of much of the Modern World. This process is intricately linked to the fact that Europeans were pioneers in the ever-changing methods of warfare leading to a 'military revolution' (a now generally accepted consensus agreed to have occurred during the fifteenth to seventeenth centuries). For survival, European states had to create complex bureaucratic structures (the fiscal-military state) in order to increasingly tax and regulate the actions of their citizens to support significant standing armies and naval forces.[13] The most successful European states of the seventeenth century and beyond were the ones that created the most efficient and participatory regimes that balanced the interests of both the individual subject and that of the state, providing the state with loyal subjects willing to pay increasingly larger taxes.

This process has a long history of evolutionary progress in England but can be said to have been given a trial by fire during the turbulent Interregnum, during which England became a European power of some significance due to its aggressive foreign policy. During the Restoration period and forward, it was in the interest of the English monarchy to continue building a robust fiscal-military state as it secured the realm and promoted its interests abroad. These goals could not be met without some degree of support from parliament and the English public at large, which in effect meant that the state sought to better balance public and private interests in a significant manner.[14] Nowhere is this process more striking than in the evolving interconnected interests of the maritime and merchant community at large, the Royal Navy, and the state itself, all of which had a vested interest in the maintenance of trade or in the predation of competitor trade along the lines of prevailing mercantilist theory.[15] These 'interconnections of interests' across broad sections of Restoration English society could occur because of the revolutionary ways in which the 'public sphere' evolved and expanded thanks to the significant growth of print culture that promoted open political discussion and the fostering of national myths.[16] Beginning in the Elizabethan age, the enduring myth of 'English sea power' was a potent political force, enabling the private maritime business enterprise of privateering to successfully find its niche in Restoration English society, as defenders of both the realm, trade and of individual freedoms. Although England's pervasive nostalgia for "Sea Dogs" such as Drake and Raleigh served to help popularize privateering enterprise, the conduct of these sixteenth century corsairs was far removed from that of the privateers of the Restoration period.[17] Whereas the corsairs of the Elizabethan age had been haphazard in their application, with scant regulation, and little to show for their efforts overall; privateering during the Restoration period and beyond became highly regulated, respectable, and much more predictable in its application and usefulness.

Privateering enterprise is thus emblematic of this reciprocal evolutionary relationship between public and private spheres in early modern England and of the generalized trend witnessed in the successful fiscal-military states of Europe, which all found ways to develop more robust bureaucratic regimes to better manage the growing scope of the state. To maximize profit and restrict unlawful action and embezzlement/fraud, the state further developed a robust and consistent regulatory apparatus, the High Court of Admiralty and Commissioners for Prizes. This apparatus regulated privateering enterprise from beginning to end, from the steps needed to own, equip, and set out a privateer ship to each and every adjudication of a lawful/unlawful prize to be divided up between investors and the Crown (which received ten percent of all proceeds). This increase in state regulation was undertaken to limit the ways in which privateering could disturb the established public order without altering the basic methods of privateering, which were rooted in medieval practice and centuries of legal precedent. This process is emblematic of the overall trend of the shift towards state monopolization of legitimate use of force and the shift from the semi-private navy of Elizabeth to a largely public state navy towards the end of the seventeenth century. This could only become possible once the apparatus of the fiscal-military state was firmly in place and the citizens of the state were ideologically brought into the fold of supporting such an apparatus in a co-operative manner and not co-opted, such as in an absolutist state.[18] Thus, the fact that private individuals in Restoration era England and beyond would willingly risk their personal interests and allow their government to regulate a private business enterprise in which they staked their own personal fortunes is quite striking and deserving of further study.

The Prosopographical Approach and Paleographical Issues

The process of reconstructing the world of Restoration privateering enterprise utilizing a prosopographical approach is fraught with difficulties due to the complex scope and surprisingly voluminous amount of surviving manuscript material spread about in various archives in both the United Kingdom and the United States. As to be expected for over 350-year-old manuscripts, many of the documents consulted contained abnormalities such as stains and mold (necessitating wearing a facemask and gloves) or were unwieldy/delicate. Therefore, digital photography of the manuscript sources was not only a necessity, but proved to be fortuitous as it facilitated the creation of a relational prosopographical database that enabled a more efficient organization, transcription, and analysis of the manuscript material.

The primary source base for this present prosopography is the voluminous loose records of the High Court of Admiralty at The National Archives (TNA) in Kew, England. The High Court of Admiralty (HCA) and Records of the High Court of Delegates (DEL) series of records at the TNA contain all the critical sources of information needed to re-create the world of privateering enterprise: warrants, affidavits, captain's declarations and most importantly the bond for each privateer ship.[19] In addition, there was a 'list of privateers' (TNA, HCA 25/210) which contained data on 42% of the 101 known privateering voyages emanating from England during the Second Anglo-Dutch War and a 'register of commissions of reprisal' (TNA, HCA 25/228) which contained data on 79% of the ships in the same category. While there are 86% of the warrants for the 101 known privateering voyages, these documents are of limited use as they contain minimal information. The other classes of documents are few in number comparably; there were only 31 captain's declarations and 16 sets of affidavits.[20]

The situation for the Third Anglo-Dutch War is similar (97% of voyages have warrants) despite the dramatic drop in the number of documented privateering voyages, from 101 to 30. Listed in the aforementioned register book (HCA 25/228) are 76% of the ships, which are now given a unique standardized number based on their departure date. There is a dramatic increase in the number of captain's declarations, (20 of 30 voyages, 67%) which now contain detailed information on the privateer ship never before documented, on both the provisions of the ship and names and ranks of officers of the crew.[21] Both of these new practices are a definite sign the state is beginning to assert more control and is more tightly regulating privateering enterprise. Keeping strict control over the victualling of these ships (as noticed by the details now contained in the captain's declarations) was certainly done to prevent the privateersmen from violating the terms of their restricted cruising schedule by supplying themselves for a longer voyage than stipulated or by having more weapons than allowed (indicating a larger crew than stated in the declaration).[22]

Of all the records which help reconstruct the business and regulation of privateering enterprise, none is more important than the bond. For all intents and purposes this was the 'letter of marque' carried by the captain as proof of his status as an officially licensed private warship operating under the strict guidelines of privateering conduct mutually agreed upon, by all the maritime nations of Europe of the time.[23] The bond not only lists the ship's name, tonnage and date the bond was signed, but critically lists detailed information on the captain and owners, giving their location by parish and their occupation or titled status. It is primarily with the information gleaned from these bonds that significant trends pertaining to the intersections of interest between the public and private sphere in England during the Restoration period were revealed. The most significant trend coming to light was the very tangible co-operation (not co-option) of privateering enterprise with the greater London merchant community (for example, charter trading companies), the Royal Navy, and even the state itself, all at a time when subjects of other states in Europe had no comparable opportunities, with the exception perhaps of England's arch rival, the Netherlands.

Naturally there are limitations to the source base available for this study; it is certainly not complete, as there are privateers mentioned in the *London Gazette*, *Orders in Council*, or *State Papers Domestic* that did not show up in the incomplete collections in TNA. Many of the documents are in poor shape or could contain errors themselves; this was borne out in the cross-checking that was possible by matching the names and stats for each ship in the various different types of sources available. In addition, I have not included the quite significant contribution of Scottish privateers in the Dutch wars nor those set out from outlying areas other than England and Wales; the former has recently been given significant attention, and scant records exist for the latter.[24]

This present study endeavours to open a window into a larger understanding of how the elites and the middling sort found mutual interests in partnering in a business venture that was regulated by the state in a revealingly modern way which resulted in mixed results; two examples will demonstrate how privateering enterprise found mutuality and common interest with other sectors of Restoration English society with beneficial outcomes. By no means could it be said that these intersections of interest always came out with positive results. A third and final example involving the legendary Samuel Pepys will demonstrate how a private business enterprise conducted by individuals with overlapping responsibilities within that of the state itself could create conflicts of interest, disrupt state policy, become a source of embezzlement (a problem equally rampant in the Royal Navy,) and cause diplomatic incidents of some significance. Ultimately though, the positives outweighed the negatives, as evidenced by the continued promotion of privateering enterprise by European states through the Napoleonic period, over a century and a half later.[25]

The Reciprocal Interests of Privateering Enterprise and the Company of Royal Adventurers of England Trading to Africa

Intersections between private and public interests in regards to England's maritime interests in the early modern period, as an evolving phenomena and trend, have roots going back to the medieval period, before there was such a thing as a 'state navy;' nascent state navies in Northern Europe were a new phenomena of the mid- to late sixteenth century.[26] We have already briefly discussed the problematic conceptualization and definition of 'privateering'; the major differences between pre- and post-interregnum privateering enterprise; and the slow rise of the state of England's monopoly of the application of violence on the high seas in correlation to the growing size and scope of the Royal Navy itself in the century preceding the Restoration. Once the undeclared war with Spain (1585-1604) ended, England was free to turn its energies to new colonial and maritime trading endeavors armed with both the newfound knowledge of the Atlantic World and deep ocean maritime expertise gained by corsairs such as Drake and Raleigh.[27] These new endeavours were accomplished primarily through state-sponsored charter companies which granted certain 'rights or privileges'. These charter companies, as with privateering enterprise, had medieval origins. For all intents and purposes the shareholders were given a monopoly to trade in a specific region in order to obtain specific valuable items that could be sold at great profit at home or in markets abroad. These regulated monopolistic charter companies would not only thwart rivals from other competing European states, but would also stop fellow individual Englishmen from cutting into the profits by going it alone (known as interloping) and thus lessen the profits of both the company and the crown itself, as the crown was typically a shareholder as well.[28]

The intersections of interests between public and private maritime interests evolved more significantly when the nature of charter companies began to assume their more recognizable 'modern' joint-stock form, which occurs for the first time during Queen Elizabeth's reign.[29] As with privateering enterprise, the main idea behind a joint-stock charter company is to enable a group of investors to band together to form a partnership to undertake a risky business venture requiring more capital than was available to any one individual and to spread the risks of the inherently dangerous seaborne maritime activity; naturally any potential windfall would be split amongst the investors according to the agreed upon terms signed in the contract which was backed by the Royal Seal.

In the sixteenth century, English state control over these joint-stock charter companies was somewhat restricted due to

Fig. 1. Samuel Pepys by Sir Godfrey Kneller, 1689. Courtesy National Maritime Museum.

the limited financial resources of the crown and the lack of regulatory agencies or established legal precedent, but as the seventeenth century progresses, a growing trend of increasing state control over these joint-stock charter companies can be discerned. This trend is emulated as well in regards to privateering enterprise, particularly after the Interregnum. It is important to note why it was crucial for a burgeoning fiscal-military state of the seventeenth century to exert more control over these joint-stock charter companies, particularly as it readily became clear that these companies often became a liability (e.g., the ill-fated experience of the Virginia Company whose charter was revoked by King James in 1624).[30] Not only could the state stand to lose its investment if there was mismanagement, but it was in the state's interest to keep the company in line, particularly in regards to foreign policy. The company was in a real sense a representative of the state overseas, and its actions could have far-reaching repercussions, as evidenced by the events of 1663 in which a charter company, the Company of Royal Adventurers of England Trading to Africa (CRAETA), single-handedly started a war. This adoption of more rigorous regulation of charter companies and privateering enterprise in England evolved concurrently with another significant trend throughout Europe: a consensus on general maritime law and precedent that evolved from the medieval 'Laws of Oleron', which were further codified succinctly by the talented Dutchman Hugo Grotius; thus, by the mid-seventeenth century a concept known as the 'Law of the Seas' was well established.[31]

In contrast to the authoritarian control of charter companies in absolutist states such as France where power was centralized and its elites co-opted or the United Netherlands with its strict decentralized public control, public state control of charter companies in England tended to be haphazard, less formal and subject to interpersonal relationships of patronage and nepotism that often did not favor the state's interests, resulting in embezzlement. The contributions from the state tended to be smaller, and little or no military support was ever provided; the companies were very much on their own.[32] This meant that ships employed by English charter companies on long voyages had to be well armed and ready to defend themselves against a host of pirates, corsairs of the Mediterranean or rival Europeans. Most would then set sail with a specialized letter of marque to give the company ship a right to defend itself, and if the ship sailed during a time of war, the right to capture a prize if an opportunity presented itself; the prize would then be adjudicated as a lawful prize by the High Court of Admiralty when the captured ship was brought back to port.[33] These merchant ships and opportunistic privateers are quite distinct from the purpose-built warship privateers (which were not engaged in long-distance trade) that were given a specialized letter of marque restricting the length of time the privateer could go cruising against a specified enemy during a time of open war. It is this latter category that is the basis of the present study.

The phenomena of a relative lack of direct state control and financial interest in the joint-stock charter companies by the English state (in comparison to the French or Dutch case) began to change during the Restoration. A new joint-stock charter company, the aforementioned CRAETA, is the best example of how the crown came to be more directly involved in the affairs of charter companies. CRAETA's first charter of December 1660 gave it the exclusive right to search for gold on the coast of Africa.[34] This company would receive special attention from the crown for a very important reason: 24 of the 32 subscribers (75%) were titled or members of the royal family; subscribers included James, Duke of York, the Princesses Maria and Henrietta, Prince Rupert, and the Dukes of Albermarle and Buckingham, among others. Coupled with the fact that James, Duke of York, (who was the director of the company) was also Lord High Admiral of the Navy and held other numerous significant positions of power meant that this 'private' joint-stock charter company (composed primarily of the courtly elites of English society) could be considered a quasi-semi-private extension of the state. CRAETA was reformulated with a second charter in 1663 which granted monopoly trading rights in western Africa for 1,000 years, specifically mentioning the slave trade for the first time and with its membership now growing to 66 subscribers. Membership in CRAETA still reflected a high percentage (61%) of titled members in addition to several new Royal family members which now included Queen Katherine and Mary, the Queen Mother.[35]

The intersections of public and private interest continue even further; 11 subscribers of CRAETA were also members of the Council of Trade created in 1660, and eight subscribers were members of the Council for Foreign Plantations.[36] The final CRAETA charter of 1667 (which now includes the King himself), issued shortly before the company collapses that very same year, continues the same trend with shareholding dominated by members of peerage rank or higher, at 68% of the 112 subscribers. As will be demonstrated later with data from the prosopography database, privateering enterprise is far less restrictive and more representative of the 'middling sort' than charter companies such as CRAETA; it points towards the future of how the fiscal-military state of England will evolve in the long eighteenth century with broad support from a widespread portion of English society, and not just the elites.[37]

There is no better example of how a charter company could affect the foreign policy of a state in a significant way than that of the cabal controlling CRAETA, led by its director James, Duke of York, which implicitly directed Sir Robert Holmes (using both Royal Navy and CRAETA ships) to raid all the Dutch outposts on the African coast in 1664, knowing full well this would lead to war.[38] That this expedition sailed in the first place demonstrates how the web of intrigue and confluence of interests among the clique of those influential members of the Royal family, the maritime and merchant community of London, and members of the Royal Navy dominated by James the Duke of York could single-handedly change the course of a state's foreign policy and lead to a declaration of war.[39] Although Holmes's expedition was an initial success, the retaliatory expedition sent by the Dutch under Michiel de Ruyter recaptured everything the Dutch had lost with the exception of the Cape Coast Castle; CRAETA had expected a windfall but in the end received very little financial compensation for all its efforts, and Holmes temporarily found himself in the Tower of London until war was officially declared.[40]

A closer examination of the subscribers from all three CRAETA charters reveals a significant link to privateering enterprise with several individual subscribers who would themselves become shareholders in a privateering consortium during the Second Anglo-Dutch War of 1664-1667. Subscriber Sir Richard Ford (a London merchant and MP from 1662-1677) was an owner of the *Sparke Friggott*, which set out in June 1667, and of the infamous *Flying Greyhound*, which set out in October 1666, of which much will be discussed in the third section of this paper.[41] The privateer *Mary of Guernsey*, which set out in January 1667, had two CRAETA subscribers in the consortium: Thomas Childe, an apothecary from St Peter's Cornhill, and John Noell, reasonably said to be one the five sons of Sir Martin Noell. Sir Martin Noell was a renowned financier, merchant, political figure and owner of 20,000 acres in Jamaica who had died in 1665 and whose name continued to be listed on the charter as 'executor of his estate'.[42]

Another interesting case is that of the *Lilly*, a privateer which set out in May 1667, with a London consortium containing three titled men and a naval officer named George Colt (Lieutenant 1663, Captain 1666, drowned 1675). This is certainly the same *Lilly* which is later a part of Henry Morgan's infamous Panama 'expedition' of 1670-71; its local affairs in Jamaica were handled by CRAETA subscriber Sir James Modyford, brother of the notorious governor of Jamaica Thomas Modyford (the *Lilly* still being owned by a London consortium despite being in the far flung Caribbean).[44]

Equally intriguing is the curious instance of an owner partaking in a privateering consortium while imprisoned in the Tower of London. Sir Thomas Modyford had turned a blind eye to privateering activity during his tenure as governor of Jamaica after peace between England and Spain had been declared in 1670. The Treaty of Madrid strictly forbade the granting of any further letters of marque, which Modyford had openly continued to do. Thus, Modyford and the infamous 'privateer' Henry Morgan (who continued to cruise against the Spanish with unlawful letters of marque making him consequently a pirate) were locked up in the Tower of London until 1674 to appease the Spanish. Eventually both were later fully pardoned, rewarded, given titles, and returned to Jamaica in 1675.[45] Apparently this imprisonment did not prevent Modyford from further intrigues. During the Third Anglo-Dutch War, the privateer *Jamaica Merchant*, of the newly chartered Royal Africa Company (a charter company built up from the ashes of CRAETA, henceforth RAC), set out in November 1672 with orders to sail to Jamaica.[46] As customary for a ship on a long non-cruising voyage during a time of war it was issued a letter of marque. The *Jamaica Merchant* carried passengers and badly needed supplies for Jamaica, and its owner consortium lists the imprisoned Sir Thomas Modyford, Sir John Kempthorne (a prominent Royal Navy officer) and Charles Modyford, among others.[47]

A privateer consortium is thus a miniature version of a joint-stock monopoly company. That there were members of CRAETA who were also individual owners in a consortium for a privateer ship is indicative of just how interlinked these business communities with maritime and mercantile interests were in Restoration England, particularly for the geographical area of London and Middlesex County. Eight of the surviving affidavits (analyzed for the prosopographical database) for eight privateering voyages of the Second Anglo-Dutch War had no direct ties in their consortiums to CRAETA but were still signed at the Africa House (Warnford Court, London), the headquarters for CRAETA and later the RAC, indicating again the interrelations between these close-knit communities involved in inherently risky maritime business ventures.[48]

Another significant feature of privateering enterprise is the employment of out-of-work or future Royal Navy officers; CRAETA was keen to pick captains with experience and expertise; four of the six captains chosen for its privateering voyages were Royal Navy officers, some with illustrious careers. CRAETA privateer ships cruised to its trading concerns off the coast of Africa (much further from the standard hunting grounds of most privateers), requiring the sending out of large ships; four of the six privateer ships sent out by CRAETA were over 100 tons, significantly larger than the average privateer ship sent out during the war.[49]

During the Second Anglo-Dutch War, CRAETA was a significant player in

Table 1
Privateer Ships of the Company of Royal Adventurers of England Trading to Africa

March 1665[6] (before war declared)

Barbadoes Merchant 230 tons
John Heath: L 1661, C 1664

Royal Catherine 30 tons*
Thomas Southing

Sampson Friggot 90 tons
Peter Edwards: L 1665, C 1672

July 1666

Charles 120 tons
John Hayward: L C 1660, KIA 11 Aug 1673

Golden Lyon 350 tons* +
Abraham Holditch: L 1661, C 1665

James 140 tons*
John Yard

L = Lieutenant. C = Captain
* Ship part of the 1664 Holmes expedition to Africa.
+ A Dutch prize from the 1664 expedition turned over to CRAETA.
Ship information: TNA, HCA 25/9; TNA, HCA 25/10
Captain information: Syrett & Charnock (see reference 52)

privateering enterprise, sending out six ships in total. The intersections and conflict of interests are quite illuminating in this case. When each of these ships set out, the warrant given to each captain to receive their Letter of Marque was signed by James, Duke of York, in his capacity as Lord High Admiral (ironically, also their boss, as director of CRAETA). In addition there is only one witness and guarantor for the bond (a special privilege no other privateer ship enjoyed) for each of these six CRAETA privateer ships: the secretary of CRAETA, Sir Ellis Leighton. Leighton himself had a conflict of interest; he was appointed to the Prize Commission Office in 1664 (established to manage the selling of goods and ships deemed lawful prizes by the High Court of Admiralty) and was appointed as one of the King's Counsel in the Admiralty court.[50] This meant that the usual security of £4,000 may not have actually been paid, and these ships were no doubt given preferential treatment both in the outports and perhaps in the High Court of Admiralty when prizes were adjudicated upon.[51] All these fortuitous advantages were of course privy only to the elite few involved in the same circles of power as those who enjoyed these privileges and had a hand in formulating state policy. In this case, the conflicts of interest, and the blatant misuse of overlapping of public and private responsibilities, clearly benefited those in both CRAETA and privateering enterprise.

Of the four Royal Navy captains employed by CRAETA, three had been promoted to captain before their privateering voyage and thus took the opportunity to gain employment while they were currently without commission. This was a common trend; a complete listing of all indentified Royal Navy captains who set out on a privateering voyage is provided in appendix 1 and 2. Of the identifiable privateer captains of the Second Anglo-Dutch War, 21 of the 87 (24%), had past or future Royal Navy experience.[52] In the case of Peter Edwards, the captain of the *Sampson Friggot*, this trend is quite apparent: after a presumably successful privateering voyage in 1665, he found at least four commissions after his voyage on the *Sampson Friggot*, the last being captain of the *Blessing Smack* in 1672.[53] John Hayward would later serve an illustrious career, with six subsequent commissions after his privateer voyage; his last was commanding the *Royal Charles*, where he was killed in action at the Battle of Texel in August 1673.[54] Abraham Holditch had participated in both the CRAETA expeditions to Africa commanded by Sir Robert Holmes, and himself became a subscriber to the later incarnation of CRAETA, the RAC. In 1671 he successfully recaptured the Cape Coast Castle for the RAC and was active in the RAC's administration until his death in 1678. Holditch fits the typical profile of a CRAETA and RAC subscriber, given the nature of his significant wealth as evidenced by his 1666 will, which lists an estate in the County of Dorset to be willed to his wife Elizabeth; his brother Jacob and/or two other kinsman to receive 'all my houses, lands and tenements that I have in the parish of Saint Buttolph Aldgate in the City of London... and my manor of Gaston in Totnes'. Holditch even left £40 to the sister of one his servants named Edith Harvey.[55]

The fortunes of CRAETA mirrored that of England towards the end of the Second Anglo-Dutch War, and by April 1667 it was virtually bankrupt and subsequently started to grant trading licenses to private individuals in return for a payment of £3 per ton of goods obtained through trade either in private ships or existing CRAETA ships sent out on personal finances. In a final desperate act, CRAETA's rights and privileges were sold off to individuals (some of them CRAETA subscribers) belonging to a new organization known as the Gambia Adventurers, who themselves would later create the RAC in September of 1672.[56] It is worth noting that apart from the aforementioned *Jamaica Merchant* the RAC did not send out any privateer ships during the Third Anglo-Dutch War. The RAC's attention was now firmly focused on a commodity far more profitable and predictable than privateering enterprise: slavery.

The Reciprocal Interests of Privateering Enterprise and the Royal Navy

We have seen how privateering enterprise found common interest with a joint-stock company such as CRAETA during the Second Anglo-Dutch War. Members of both the elite clique who created CRAETA and the maritime community at large (comprising a significant proportion of middling sort) who were engaged in privateering enterprise found this partnership fruitful, and in this specific case it worked to their mutual benefit, despite potential conflicts of interests due to overlapping responsibilities.

This beneficial reciprocity stems from the fact that both joint-stock companies and privateering enterprise hinge on risk and speculation. One reason that consortia were such a necessary element of privateering enterprise was due to the inherent risks involved in any maritime venture; a consortium's partnership of at least two owners, most commonly three, and rarely more than three, helped spread out the risk. These risks were naturally amplified during a time of war, and any of the following outcomes could be disastrous to those who staked their personal fortunes: (a) the privateer ship could be lost at sea, either destroyed by enemy action, captured, or lost to the elements; (b) the privateer could act inappropriately or turn pirate and the owners held liable and the £4000 bond forfeit, almost unheard of in European waters; or (c) the privateer could simply return without capturing any ships, which would entail a financial loss for the consortium. Even worse, a privateer could have a prize adjudicated upon by the High Court of Admiralty ruled an invalid capture, meaning that not only were the prize ship and goods returned to the original owner, but the consortium must also pay the costs of the court and damages to the original owners.

Sailing on the open seas in the early modern period was always inherently dangerous. The privateer *Victory*, one of the most successful privateers of 1666, was lost at sea in December of that year: 'Several vessels have been wrecked on Portland beach, especially the Victory, the Duke of Richmond's privateer of 10 guns, under Captain Lucy, who was drowned with nine of his company'.[57] Simple bad luck could also be a factor even when legitimate prizes had been captured and were about to brought in to be adjudicated upon; on 15 January 1666[7] it was reported from Bruges in the *London Gazette*:

> The storms have been lately very violent upon oar Coasts as well as in other parts, by which three considerable prizes taken by our privateers, were lately cast away in Ostend Road, and two others near Dunkirk., besides many other ships, which have perisht upon several parts of these Coasts.[58]

It was for these reasons that many consortia turned to established mariners and particularly Royal Navy officers to captain their privateer ships; 21 of the 87 (roughly 24%) of the identifiable privateer captains of the Second Anglo-Dutch War had Royal Navy experience (previous service before the privateering voyage or future Navy service after the voyage, no doubt spurred on by their privateering experience). This percentage is continued for the Third Anglo-Dutch Naval War; eight of the 30 (roughly 27%) identifiable privateer captains have Royal Navy experience.[59] Privateering could provide employment to Royal Navy officers currently without a commission, a trend evident with two of the CRAETA privateer captains. Another example would be Richard Acton, a gentleman from St Giles in the Fields parish, who captained the *Byfrons Friggot* in June of 1665; he had been appointed a Lieutenant in the navy in 1661, but had not seen a commission since.[60]

More common though was the use of a privateering voyage to obtain fame and later obtain a commission in the navy, and the majority of the privateer captains with Royal Navy service in both wars had navy service after their privateering voyage. This means that privateering was not only an acceptable method of getting noticed, but was seen as a proper training ground for future navy officers of men of either titled status or from the middling sort. Of the 21 privateer captains with past or future Navy service in the Second Anglo-Dutch War, seven are gentlemen and one is an esquire, leaving the rest to be mariners, thus representing the middling sort. For the Third Anglo-Dutch War only two of the eight with Navy service are titled. These figures are reflective of the growing tensions and controversies in the Royal Navy over the 'Gentlemen and Tarpulins' debate (who best should be an officer: a gentleman with little to no experience or an experienced mariner without a title), which was perhaps the most significant issue at the heart of the very nature of the Royal Navy itself during the Restoration. The eventual resolution of this crisis had ramifications on the evolution of the professionalization of the Royal Navy from the Restoration forward.[61] Before 1677, when it became mandatory for all candidates for the rank of lieutenant in the Navy to prove a minimum service of three years at sea and pass a qualifying examination, there was no formal structure

Fig. 2. Sir Frescheville Holles (on the left) and Sir Robert Holmes, by Sir Peter Lely c. 1672. Courtesy National Maritime Museum.

of rank and no established tradition of recruiting officers from the nobility and gentry in the Royal Navy.[62] This perhaps explains why privateering enterprise played such an important intermediary role in training seamen and officers during this transition period to the exam system. Even after the exam system was fully in place, privateering service remained universally recognized as a training ground for future seamen, and throughout the long eighteenth century, privateers continued in their role as an important ancillary force that had a significant military, strategic, and economic utility which was provided at no cost to the state.

A classic example of how an aspiring man could use privateering in order to obtain a commission in the Royal Navy is Sir Frescheville Holles, esquire of Carlton, County Nottingham, who captained the *Panther* at the very beginning of the war (obtaining a letter of marque before war was declared in February 1664[5].) After a successful privateering cruise, he obtained a commission in June 1666 and went on to have a distinguished Royal Navy career, was elected an MP and became a favorite at the court. He was well renowned for his bravery, having lost his arm in the Four Days' Fight; his distinguished career was cut short when he was killed at the battle of Solebay.[63]

As noted earlier, joint-stock trading companies were primarily composed of cliques of elite society, and certainly the Restoration Navy itself was less open to the middling sort due to the unprecedented attention of King Charles II and James, Duke of York, and their strong preference for choosing noblemen and gentlemen for officers in order to create a staunchly royalist officer corps.[64] In stark contrast, the statistics from the prosopography database clearly demonstrate that privateering enterprise was inclusive and representative of a much broader section of the 'middling sort' which itself was growing in influence throughout the maritime community and beyond in England. Of the 87 identifiable captains in the Second Anglo-Dutch War, 52 list 'mariner' as their occupation, thus 60% of all captains in this war are of the middling sort. This trend continues even more so in the Third Anglo-Dutch War: 16 of 24 identifiable captains (67%) are mariners.[65] An example of a middling sort captain who moved his way up to Navy service from a privateering voyage is Robert Jones, of St Dunstans in the East, who set out in the privateer *Robert* before war was declared in February of 1664[5]. He obtained a commission in 1667 and was appointed commander of the *John and James*.[66]

Another remarkable example of a middling sort captain (and part-time Royal Navy officer) who himself was an entrepreneur extraordinaire was Thomas

Hendra, of St Marie and Martins London, who was given a captain's commission and whose ship the *Eagle* was one of the few private merchantmen hired to serve the Royal Navy in 1665 as it was quite a large ship; at 500 tons, 38 guns, it was the largest ship to serve as a privateer in the Second Anglo-Dutch War. It sailed as a privateer in May 1666 (Hendra as captain and an owner in the consortium) after it had been released from naval service. The *Eagle* with Hendra as captain continued its service to the state in the Third Anglo-Dutch War and was known to have fought in the rear division of the red squadron of the Duke of York at the Battle of Lowestoft on June 3, 1672.[67]

It is apparent that a strong case can be made that privateering enterprise served as both an outlet for unemployed skilled navy officers and seamen *and* as a training ground for future Navy officers and seamen, thus providing a significant contribution to the professionalization of the Royal Navy. One way we know this to be true is the fact that when there was a desperate shortage of available seamen for Navy ships, privateer commissions would be withheld and privateers forbidden to sail for fear of losing too many skilled seamen to these private entrepreneurial warships. There were good reasons why sailors would prefer to serve on a privateer rather than a navy ship: discipline was looser, the chance of obtaining a prize was far greater, and the amount of prize money the crew could expect from a captured ship would be higher than what they would receive for a ship captured by a Navy ship.[68] It was certainly for these reasons that, due to a chronic crew shortage, no privateers were allowed to sail from December 1665 until May 1666 and those privateers at sea before December 1665 were mandated to return no later than March 20, 1665[6]. Care was even taken to ensure merchant ships with letters of marque which sailed on longer trading voyages, 'letter of marque ships', provide a certificate to ensure compliance.[69] It was also standard practice for Navy ships in times of dire crew shortages to seize men off privateers either on the open sea or the moment the privateer left port as evidenced by this dispatch from 27 June 1666:

To the Capt. of the Colchester. We understand that there are certain Privateers ready to get out of Dover, which are well supplied with men, whilst the King's fleet is in want enough, wherefore we desire you to ply twixt that place and Calais, and strictly observe their motion, and as soon as they come out to go on board and take all the able seamen out of them. Be careful you meddle not with them whilst they are on shore and keep our Instructions privately till you have obeyed them.[70]

The dramatic drop in the number of privateering voyages during the Third Anglo-Dutch War can certainly be attributed to the chronic shortage of seamen; there was little

Table 2
Residence of Captains in the Second and Third Anglo-Dutch Wars

Second Anglo-Dutch War	Third Anglo-Dutch War
47 captains (of 65 with location info. since part of consortium) from Mddx London area, or 72%	18 captains (of 24 with location since part of consortium) from Mddx London area, or 75%
11 St Margaret's Westminster 17% 9 Stebonheath/Stepney/Wapping 14% 7 St Martin in the Fields 11% (Top 3 parishes in Middlesex 42% of captains)	6 Stebonheath/Stepney/Wapping 33%
18 captains outside Mddx London area 28%	6 captains outside Mddx London Area 25%
Source: TNA, HCA 25-9/10	Source: TNA, DEL 2-108

enthusiasm for the war and the press gangs were unsuccessful in obtaining enough seamen, perhaps explaining why there were only 30 privateering voyages in the Third Anglo-Dutch War compared to 101 in the previous war.[71]

The majority of privateer captains (75% for the Second Anglo-Dutch War and 80% for the Third Anglo-Dutch War) were themselves shareholders and owned a stake in the venture.[72] This meant that these enterprising captains, even those of the middling sort, had the wherewithal to pay their portion of the £4,000 bond. This no doubt encouraged the captain to strive for a successful voyage and capture prizes since he too had such a large personal stake in this maritime predation business venture and had good reason to act lawfully. There are several instances of Royal Navy officers owning shares in privateering consortia while not commanding them. Sir John Kempthorne (who rose to rear admiral) was a shareholder in the consortium for the *Jamaica Merchant* as was discussed earlier; George Colt [Lieutenant 1663, Captain 1666, drowned 1675] was a shareholder of the privateer *Lilly* which set out in May 1667; and Charles Whittington [Lieutenant 1672] had shares in two ships that set out in 1673, the *Guift* and the *Norfolk Friggot*.[73]

Government officials also could not resist the potential rewards of privateering enterprise; one such representative example was Colonel John Strode, Governor of Dover Castle, who had shares in two privateers during the Third Anglo-Dutch War, the *Dover Friggot* and the *Dover Castle Yacht*.[74] An even more infamous example of this kind involving none other than Samuel Pepys, Clerk of the Acts, will be discussed in the next section of this paper. It is easy to see how the Royal Navy and officials of the state saw privateering enterprise as an enticing and legitimate business venture that not only served a patriotic duty to participate in the nation's defence and attack the enemy's commerce (and will to fight), but also gave them a personal chance for a potential financial windfall.

Another significant trend worth mentioning is the primacy of London and Middlesex county environs listed as residences for both captains and consortium owners in privateering enterprise. This trend can be partially explained by the historical importance of London in the mercantile community as a major entrepôt of trade and the fact that London was the seat of government. Yet it also reflects an increase of social mobility in Restoration English society that attracted entrepreneurs of both titled status, and in increasingly larger numbers, the middling sort, to London to make their fortunes. The increasing power and scope of the middling sort mercantile community of London, as evidenced by their exuberant participation in privateering enterprise, is one of the many ways the middling sort would play a crucial role as willing financial contributors to the English fiscal-military state whose role over time started to become more significant than the traditional landed titled elites that compromised the vast majority of subscribers of charter companies such as CRAETA.

These aforementioned trends are evident among the members of privateer consortia as a whole (259 total individual owners for the Second Anglo-Dutch War and 101 individual owners for the Third Anglo-Dutch War.) The middling sort make up 53% of the 259 total owners and 68% of the 101 total owners respectively; another indication of the growing breadth of participation in privateering enterprise and the slow rise of importance in English society in general of men with such diverse occupations as an embroiderer, haberdashers, distillers, cheesemongers, chandlers, and even a theology professor, all of whom, by whatever means, came up with their share of the £4,000 bond. As with the privateer captains, these consortium owners came predominantly from the London Middlesex area: 71% of owners for the Second Anglo-Dutch War and 73% of owners for the Third Anglo-Dutch War. A noteworthy number of these entrepreneurs were owners in more than one privateering consortium; in the Second Anglo-Dutch War there were four consortia that had two of the three owners partner on other ships. One of these two-men partnerships which owned three ships was an interesting mix of a middling sort man and a titled man: The *Adventure*, *Gerard*, and *Robert*, all set out in the first two years of the war, were jointly owned by Agmondesham Pickeyes, a goldsmith from St Leonards Fosterlane Middlesex, and John Titus, esquire of Deale, Kent. During the Third Anglo-Dutch War several notable merchants of the middling sort dominated privateering enterprise. A prominent example would be James Littleton of Catherine Church London, who had shares in no less than *seven* ships, or 23% of all privateers sent out in the entire war: *Revenge* (twice), *Benjamin*, *Little James*, *Spyes*, *Resolution*, and the *St George*. Parish ties and family ties also proved to be an important factor in the formation of these privateering consortia; there were six consortia in the Second Anglo-Dutch War and three consortia in the Third Anglo-Dutch War that had two or more members of the same family as owners.[75]

What can be discerned from all of this information fleshed out of the prosopography database is the fact that privateering enterprise was quite inclusive and representative of both titled and non-titled portion of society in Restoration Britain – particularly for those participating from Middlesex County – and served as both an outlet for unemployed skilled mariners (including Royal Navy officers) and as a training ground for those who wished to gain recognition and the experience needed to obtain a commission in the navy. Having provided two in-depth examples of how the intersections of interest benefited a broad spectrum of individuals when they chose to involve themselves in privateering enterprise, we shall now explore the fact that not all instances of intersections of interest involving privateering enterprise during the Restoration period ended up with beneficial results for either a particular individual or the state.

An Innocent Misappropriation? Mr. Pepys and the Conflict of Interests Between Privateering Enterprise and the State

It is clear that while privateering enterprise found common interest with both the elites, joint-stock charter trading companies such as CRAETA, the mercantile community at large (providing opportunities for the 'middling sort' unavailable elsewhere), and provided employment and training for seamen/officers of the Royal Navy of Restoration England, it could also cause conflicts of interests that could have unfortunate results or create situations with unfavourable outcomes for both individuals and the state. One such example of this phenomenon is the surprisingly little-known controversial foray of the infamous Samuel Pepys into privateering enterprise. Pepys' career is well documented thanks to the voluminous archival sources stemming from his position in the Admiralty, his own vast personal library bequeathed to Magdalene College, and perhaps most famously, from his diary penned from 1660-1669.[76] As had been discussed earlier in our first example, the Duke of York's overlapping offices and duties seemed to be of benefit to him and those of his clique in regard to CRAETA and privateering enterprise; no significant problems due to overlapping responsibilities or conflicts of interest occurred in that instance. In Samuel Pepys' case, as Clerk of the Acts, he was responsible for numerous important administrative tasks for the state Navy and was deeply personally connected to many important personalities at Whitehall, including the Duke of York himself. In this next case example, overlapping responsibilities and intersections of interests would prove to cause a serious diplomatic incident for the English state and almost end Pepys' career; only by chicanery, bribery, and luck did both Pepys and the state come out of this misadventure of top Navy officials forming a privateer consortium relatively unscathed. Despite repeated ill-advised high stake gambles made by Pepys and his high-profile privateering consortium, Pepys' venture into privateering enterprise would ironically leave him and his consortium partners significantly richer.

The fact that a member of the Navy would involve himself in a privateering consortium (or captain a privateer himself) was commonplace, but to have men who were top level administrators of the state Navy get involved in a private business venture which could create significant conflicts of interests in regards to state policy and naval matters is perhaps surprising, even for Pepys' day. The reason why Pepys sought to partake in privateering enterprise was simple enough; despite powerful patrons and sweeping authority over naval matters, Pepys was ever vigilant for opportunities to supplement his income, which is surprising considering for his day his salary was no small sum for an untitled government servant; £350 per annum.[77]

One such opportunity for financial advancement arose in September of 1666, when Pepys contrived with Sir William Penn (an Admiral and Commissioner of the Navy Board) and Sir William Batten (Surveyor of the Navy and MP for Rochester) to become joint owners of the privateer, the *Flying Greyhound*. Pepys' diary entry on 26 September 1666 certainly expresses the desire of all men, regardless of their prominence in society or apparent wealth, who sought to increase their wealth by joining the boom in privateering enterprise:

Away with Sir W. Pen...We did also discourse about our Privateer, and hope well of that also, without much hazard, as, if God blesses us, I hope we shall do pretty well toward getting a penny.[78]

The *Flying Greyhound* did indeed live up to its peculiar name and prove to have an interesting and illustrious career; almost bringing down the careers of four of the most prominent men in Restoration English society. The *Flying Greyhound* first becomes known to us as having been captured by the Royal Navy ship *Pembroke* commanded by Richard Goodlad at some point during late summer of 1666, and first appears in dispatches in August 1666 with numerous accounts of its repair estimates, subsequent repairs and later convoy duties protecting colliers.[79] Pepys and his consortium partners at some point in September successfully petition King Charles II himself to have the *Flying Greyhound* lent to them at no charge to set out as a privateer on the condition that it return to the King's service as a fireship the following spring; paying nothing for the loan of the ship was most certainly a reward for Pepys' dutiful personal service to the Duke of York. But Pepys had more designs to maximize profits and minimize costs with methods utilizing state resources that no other privateering consortium could possibly employ – the dubious nature of which was beyond doubt. The first act of subterfuge was to remove his name from the privateering consortium, and thus remove himself from responsibility and liability for the conduct of the privateer as his name does not appear on the bond signed on 26 October 1666 (see appendix for the bond).[80] This was accomplished by adding a fourth member to the consortium who would put his name on the bond in place of Pepys; this was Sir Richard Ford (himself a prominent member of CRAETA). This deal had been arranged three days before:

At noon Sir W. Batten told me Sir Richard Ford would accept of one-third of my profit of our private man-of-war, and bear one-third of the charge, and be bound in the Admiralty, so I shall be excused being bound, which I like mightily of, and did draw up a writing, as well as I could, to that purpose and signed and sealed it.[81]

It was in this way that Pepys would hide his involvement in the consortium should trouble arise. Then there was the issue of how Pepys and his consortium partners would pay the crew of the *Flying Greyhound*. Most privateersmen in the Anglo-Dutch Naval Wars sailed without regular wages gambling on the shares they would receive from captured prizes, shares notably much higher than any Royal Navy sailor could expect to receive – hence the popularity of privateer service over that of the Royal Navy.[82] Those who sailed on prolonged cruising voyages however, were sometimes given wages, and it appears in this case perhaps to keep the crew of the *Flying Greyhound* happy, it was decided to pay them wages since the ship went out on repeated cruises from November of 1666 to September of 1667.[83] Pepys was thus in a precarious position of a personal conflict of interest between his private interests in privateering enterprise and his official capacities for the state Navy. Pepys chose to abuse his power of office and do something no other privateer consortium could dream of doing; using state funds to pay private employees. He therefore carefully diverted state money through a practice known as ticket-broking (buying a sailor's pay ticket at a discount and cashing it in full at the ticket office) to pay his privateer crew, at the expense of honoring payment tickets to men who were actually serving on the King's ships.[84] Pepys admits in his diary how haphazard the ticket office was (under his supervision) and perhaps it was his hope that this fact would hide his embezzlement of state funds:

> *I with all my clerks and Carcasse and Whitfield to the ticket-office, there to be informed in the method and disorder of the office, which I find infinite great, of infinite concernment to be mended, and did spend till 12 at night to my great satisfaction, it being a point of our office I was wholly unacquainted in.*[85]

When Pepys will later be accused and made to account for his embezzlement of state funds in April of 1668, it is his own clerk James Carcasse who will be his chief accuser; we will return to the ticket scandal and its resolution after we discuss another controversy involving a conflict of interest between privateering enterprise and the state as borne out by Pepys' by now legendary ship.

Meanwhile the *Flying Greyhound*, commanded by Captain Hogg, was spectacularly successful in its cruising voyages; it captured a ship from Lübeck on 22 December 1666, and later two more prizes on 29 December 1666 worth £4000, with Pepys noting in his diary 'yet, blessed be God! and I pray God make me thankful for it, I do find myself worth in money, all good, above £6,200; which is above £1800 more than I was the last year'. On 7 January Hogg brought in two more ships which appear to be rich prizes as well, but there turns out to be a major problem; the ships captured are Swedish – a neutral nation and in fact England's only true ally at this point during the war.[87] Thus, we have a classic example of a private warship authorized and regulated by the state causing a diplomatic incident; a case where private interests, utilized by the state, have become a liability to the state. An examination of the *Orders in Council* for the years 1665-1667 are rife with examples of Swedish ships being brought into English outports as prize to be adjudicated by the High Court of Admiralty and then subsequently let go. The adjudication process for these ships were closely watched by the Swedish Resident, Baron Leijonbergh, who tirelessly advocated for their release and demanded compensation to be awarded if the Swedish ship had been found to be wrongly seized, which was often the case. One such representative example from the Orders in Council from September of 1666 demonstrates this all-too-common occurrence: 'Discharge of two Swedish ships captured by the privateer *Fanfan*, the *St. Peter* and *St. Jacob* satisfaction to be given to them all goods returned'.[88]

The man in charge of High Court of the Admiralty, the highly esteemed and masterful purveyor of maritime law respected by his peers and foes alike, Sir Leoline Jenkins, adjudicated many cases involving neutral Swedish ships brought into English ports during the Second and Third Anglo-Dutch Wars.[89] Not every ship purporting to be Swedish was in fact so, as it was a common ploy for the Dutch or other enemies of England to sail under false pretenses in order to evade capture; complete with fake passports, flags and sea-briefs.[90]

A wonderful collection of reports written by Jenkins preserved in The National Archives gives a glimpse into just how complicated and intricate the rules and procedures involved in the adjudication of a captured ship brought in to an outport to be condemned as lawful prize were; the choice of whether to condemn the prize or return the ship to the original owners could certainly be a very difficult in certain cases. Two examples from this collection of adjudication reports will give us an idea of what Pepys and his consortium partners likely faced when their controversial Swedish prizes were adjudicated upon by Judge Jenkins.

As with the majority of these cases, the Swedish Resident was vocal in his advocacy for the rights of his fellow Swedish nationals. In this first instance, on 31 October 1665, Jenkins is reviewing the case of the ship the

Golden Lamb in which he notes:
> There is no Sea Brief, passport, or bill of lading among her papers....One witness more (a Holsetiner) says that the Captain appears to be a Frieslander...Unless your Lords. finding what is suggested (that this ship was violently seized & carryed away from a free port, where she lay taking in her lading and that she was fouly pillaged. that her materiall documents are by the taker subducted and not brought in) to deserve credit yr Lds shall resolve upon as more compendious way than of a process to give satisfaction to the Resident.[91]

In other words, the English privateer captain who captured this Swedish ship, while at port, (itself an unsightly practice) most likely illegally threw overboard the legitimate passes that would have proved the ship to be Swedish. In addition, the only witness making a seemingly 'damming' statement that the captain of the *Golden Lamb* is Dutch cannot be trusted (being from Holstein, a Dane, hardly a reliable witness.) Thus, in a case such as this, Jenkins had to err on the side of caution and ultimately not let this Swedish ship be condemned as lawful prize. No systematic study has been done as of yet of the prize court sentences located in The National Archives (partly because of their damaged state and that they are written in barely legible Latin) for the Anglo-Dutch Wars of the Restoration to verify the quite plausible claims made by several historians that upwards of one half to two-thirds of prizes brought in to be adjudicated upon were not deemed lawful prize.[92]

In another case (adjudicated on the same date as the *Golden Lamb*) there is a markedly different outcome. In this instance, a ship called the *Sun of Riga* was caught by an English privateer with "valid" documents (passes and sea-briefs) declaring that the ship and goods belonged to Swedish subjects living in the city of Riga, then part of the Swedish Empire. Closer examination of the facts revealed that the captain of the *Sun of Riga* and his son (who was also on board) contradicted each other in their depositions, and several crewmen speak Dutch. Irrefutable evidence in support of the validity of the prize claim for the English privateer was the *Sun of Riga*'s bill of lading which listed numerous contraband items for one 'Sebastian Mull' in Holland. In his usual humble tone, Jenkins 'submits his suspicions and presumptions humbly to your lordships'.[93] Thus, in this obvious case of subterfuge, the ship was condemned and the privateer crew and consortium owners given their fair share of the spoils while the crown received its customary 10 % of the proceeds.

It is in the context of the recent problematic relations with England's only tangible ally and the complex legalistic world of maritime prize law that Pepys and his consortium partners were to have their day in the High Court of Admiralty. It seems from the outset it was going to be a complicated trial in regards to whether Captain Hogg of *Flying Greyhound* had acted appropriately in his dubious seizure of two neutral Swedish ships. Baron Leijonbergh once again rose to the occasion to defend the interests of his fellow countrymen, and in this instance he is particularly aggrieved as the consortium owners of the *Flying Greyhound* are themselves all in the upper echelons of the English State itself and are thus responsible for a potentially significant breach in the alliance between England and Sweden. It is also important to reiterate that this particular incident comes after the long string of seizures of neutral Swedish ships and that the ships seized by Captain Hogg, (one of them called the *Phoenix*) are richly laden. Pepys and his consortium partners once again solve the problem with subterfuge worthy of Machiavelli himself:

> and then to Sir W. Batten's, where [Sir] W. Pen, [Sir] R. Ford, and I to hear a proposition [Sir] R. Ford was to acquaint us with from the Swedes Embassador, in manner of saying, that for money he might be got to our side and relinquish the trouble he may give us. Sir W. Pen did make a long simple declaration of his resolution to give nothing to deceive any poor man of what was his right by law, but ended in doing whatever anybody else would, and we did commission Sir R. Ford to give promise of not beyond L350 to him and his Secretary, in case they did not oppose us in the Phoenix (the net profits of which, as [Sir] R. Ford

cast up before us, the Admiral's tenths, and ship's thirds, and other charges all cleared, will amount to L3,000) and that we did gain her.[94]

In effect, with the Swedish Resident now bribed and out of the picture, the chances that the *Phoenix* and the other Swedish ship will be adjudicated as lawful prizes for Pepys' consortium have vastly increased. The trial concluded on 21 March 1666[7]; let us allow Pepys describe the outcome:

> *by, in the evening, comes Sir W. Batten's Mingo to me to pray me to come to his master and Sir Richard Ford, who have very ill news to tell me. I knew what it was, it was about our trial for a good prize to-day, "The Phoenix," a worth two or L3000. I went to them, where they told me with much trouble how they had sped, being cast and sentenced to make great reparation for what we had embezzled, and they did it so well that I was much troubled at it, when by and by Sir W. Batten asked me whether I was mortified enough, and told me we had got the day, which was mighty welcome news to me and us all. But it is pretty to see what money will do. Yesterday, Walker was mighty cold on our behalf, till Sir W. Batten promised him, if we sped in this business of the goods, a coach; and if at the next trial we sped for the ship, we would give him a pair of horses. And he hath strove for us today like a prince, though the Swedes' Agent was there with all the vehemence he could to save the goods, but yet we carried it against him.*[95]

Thus, even before the adjudication had been finished, goods had been embezzled and sold (which was strictly prohibited), more people had been bribed (one agent with a pair of horses) and the Swedish agents were powerless to stop the shameless condemnation of these ships because the Swedish Resident was not using his influence in this case as 'it is pretty to see what money will do!'

Pepys and his consortium partners certainly dodged a bullet with the *Phoenix* scandal (and actually came out of it richer) but this case is an emblematic example of how privateering enterprise could be a liability to the state's strategic and

Fig. 3. The Dutch flute St John Baptist by Cornelis Pietersz Mooy, 1666. Courtesy National Maritime Museum.

diplomatic interests and is also an example of how rife embezzlement, bribery and backroom deals subverted due process in not only the High Court of Admiralty, but throughout the Royal Navy itself, the state and beyond. It should be noted that embezzlement and other cases of mismanagement were chronic problems throughout the Royal Navy and one need look no further than the infamous 'Prize Goods Scandal of 1666' involving several prominent admirals who embezzled cargo from two rich East India prizes captured after the disastrous Bergen raid; it temporarily ended the career of Admiral Lord Sandwich.[96]

Back to our intrepid Captain Hogg and the *Flying Greyhound*, which continued to provide both service to the state by providing convoy protection for colliers in April of 1667 and continued to bring in valuable prizes for its owners.[97] Fortune favours the *Flying Greyhound* once again as she captures the *St John Baptist* laden with valuable Canary wine worth £4962.[98] This turn of good fortune paradoxically starts to pull apart Pepys' own consortium from within as the consortium partners begin to conclude their affairs in acrimony with the approach of peace being finalized with the Dutch in July of 1667. Pepys notes quite hypocritically on 25 July: 'it appears that Hogg is the eeriest rogue, the most onservable embezzler, that ever was known'.[99] In the end, Pepys not only fell out with Hogg, but also with the widow of Sir William Batten (Sir William having died in October 1667). Sir William died with significant debts and despite Elizabeth Batten's distress Pepys demanded his share of the £666 Batten owed him from the *Flying Greyhound* prizes. Ironically, Elizabeth Batten perhaps gets the last word when she later marries the previously bribed Swedish Resident, Baron Leijonbergh.[100]

It is not until March 1667[8], that the disgruntled former clerk of Pepys, James Carcasse, presents his case against his former employer and it appears Pepys is going to be held accountable for his embezzlement regarding the *Flying Greyhound*; yet Pepys is both confident and nervous:

> *by coach to Sir W. Coventry's; and there, largely carrying with me all my notes and papers, did run over our whole defence in the business of tickets, in order to the answering the House on Thursday next; and I do think, unless they be set without reason to ruin us, we shall make a good defence ... Hither comes Carcasse to me about business, and there did confess to me of his own accord his having heretofore discovered as a complaint against Sir W. Batten, Sir W. Pen and me that we did prefer the paying of some men to man "The Flying Greyhound" to others, by order under our hands. The thing upon recollection I believe is true, and do hope no great matter can be made of it, but yet I would be glad to have my name out of it, which I shall labour to do; in the mean time it weighs as a new trouble on my mind, and did trouble me all night.*[101]

Finally his day of reckoning comes on 20 April 1667; there is no record as to the events of the proceedings but Pepys is defiant before his arrival:

> *Up and busy about answer to Committee of Accounts this morning about several questions which vexed me though in none I have reason to be troubled. But the business of The Flying Greyhound begins to find me some care, though in that I am wholly void of blame.*[102]

In the end, Pepys (and his consortium partners) remained unscathed, despite their numerous dubious activities, and have all become significantly richer thanks to the exploits of the *Flying Greyhound*. Yet all of these incidents serve as stark examples of how privateering enterprise could cause conflict of interests that could be detrimental to both individuals and the state, and although in the end little permanent damage was done in this particular case, this kind of behaviour could potentially be ruinous for the viability of all parties involved. Privateering enterprise, the merchant/maritime community at large, the Royal Navy, the efficacy of England as a fiscal-military state, and lastly the prestige and diplomatic standing of England as a whole, could all be adversely affected by unscrupulous behaviour within a particular privateering consortium. Had Pepys and his consortium partners followed the strict rules set down by their very own offices and followed the regulations as dictated by the state they could have avoided much of the trouble they got themselves embroiled in. It again demonstrates how just how pervasive and how enticing privateering enterprise was in the seventeenth century; Pepys could just not refuse the enticement of partaking in what was an established and respectable business venture with centuries of precedent; and he was certainly not alone—privateering enterprise's popularity and utility would continue long into the eighteenth century and beyond.

We cannot leave this episode without detailing the ultimate fate of the intrepid *Flying Greyhound*; she met her end just four years later while serving as a merchantman but not without a fight – on 28 October 1671, she was attacked by two Algerine Corsairs and fought them off courageously until the ship sank earning the respect of their foes, 'the Captain dying four days after of his wounds notwithstanding all the care of the Algier Captains to preservd his life'.[103]

Conclusion

It is during the second half of the seventeenth century that privateering reached its final modern form: privateers were no longer a private *substitute* for public state navies as had been the practice of employing corsairs with individual letters of reprisal during the Elizabethan Era; privateers were now a vital strategic and economic *supplement* to the already stretched and costly battlefleets of the period, an ancillary force that could harness the energies of the private sector for state ends to the mutual benefit of both. The 'intersections of interest' between privateering enterprise, charter companies, the merchant community at large, the Navy, and the state as a whole during the Restoration period enabled a patriotic and sometimes profitable business enterprise (serving both public and private interests) to develop and evolve in relation to the growing fiscal-military state apparatus in England. The formalisation and juridification of the relationship between privateers and the English state during the Restoration underscores the fact that the shift from a private to a public navy was no automatic consequence of the rising wealth of the state, and points to the persistent influence of 'private' elites and middling sort in both matters of state and maritime\merchant ventures.

It is remarkable how a complex and multifaceted social spectrum across Restoration England willingly sought out partners to jointly stake their fortunes for the greater public good during a time of war (with a few making a profit); the *mutuality of interest* between public/state and private spheres engaged in privateering enterprise in most cases brought benefits to both spheres – as opposed to monopoly charter companies that more often than not upset the public order. This reciprocal relationship of privateering enterprise within English society demonstrates a willing co-operation (not co-option as in an absolutist state) of elites and a broad cross section of the middling sort necessary for a burgeoning fiscal-military state. The scale of public investment in the state is considerable during the Restoration period and beyond; individuals, particularly those in the maritime and merchant communities (many of whom joined together in privateering enterprise), have willingly bought into the ideological and financial needs of the state. In addition, privateering enterprise served as both an outlet for unemployed skilled mariners (including Royal Navy officers) and as a training ground for future Royal Navy officers; its significant contribution to the professionalisation of the Royal Navy for officers and seamen alike and its utility as an ancillary force to the state Navy meant that privateering (despite being borne out of medieval practice) survived well into the nineteenth century.

It is no longer tenable for historians to ignore the role and utilization of private interests during this evolutionary shifting process from private to public, which witnessed the emergence of the apparatus of the modern public state that fostered England's ambitions. Officially authorized non-state actors such as privateers as private warships during the Restoration period were, in a sense, the midwives of the fully state-controlled Royal Navy, with which England built its world empire.

Appendix I
Privateer Captains With Royal Navy Service 1665-1667

21 of 87 captains have Royal Navy service or 24%. (The 4 CRAETA privateer captains with navy service listed in previous section).

Ship	Captain	Service
Byfrons Frigott	Richard Acton (Gent)	L 1661
Penelope of Dover	George Canning (Gent)	L 1668 CA 1672 KIA 28 Oct 1677
Albermarle Frigott	Henry Clarke*	L 1665 CA 1666
Peter	Thomas Diamond (Gent)	L CA 1660
Eagle	Thomas Hendra	L CA 1665
Success	Richard Hill	L 1678
Panther	Freschville Holles (Esq)	L CA 1666 KIA 28 May 1672
Rosebush	Henry Hunt	L 1672
Robert	Robert Jones	L CA 1667
Panther	Richard Keigwin (Gent)	L 1665 CA 1672 KIA 21 June 1690
Barbadoes Merchant	John Norbrooke	L CA 1665
Lilly	Thomas Page (Gent)	L 1668 CA1670
Swiftfoote	Robert Rogers	L 1678
Barbara of Dartmouth	George Sparke**	L 1678
Saint George	Thomas Thompson**	L CA 1672
Royall Dutchess	John Tozer (Gent)	L 1673 CA 1674
Saint John	John Whiston (Gent)	L 1673 CA 1677

L = Lieutenant. CA = Captain.
* Captain with Royal Navy service not an owner on that voyage (1) of (17).
** No bond for that ship unable to ascertain owner status (2).

Ship Information: TNA, DEL 2/108.
Captain Information: Syrett & Charnock (see footnote 41).

Appendix II
Privateer Captains With Royal Navy Service 1672-1674

8 of 30 captains have Royal Navy service or 27%
(Compared with the Second Anglo-Dutch War figure 21 of 87captains at 24%)

Ship	Captain	Service
Revenge	Aden, John (Adden)	L CA 1667
Have att all	Hurst, Owen (Gent)	CA 1671
Little James	Nevill, John (Nevell)	L 1673 CA 1681 Rab Casmr 1693
Betty Friggot	Pottinger, Edward (Gent)	L CA 1690 Drown 9 Oct 1690
Achilles Frigott	Potts, Thomas	L 1666
St George	Shafto, William	L 1666
Norfolk Friggot	Sparrow, John*	L 1665
Hopewell	Tildesley, Thomas*	L 1696

L = Lieutenant. CA = Captain.
* Captain with Royal Navy service not an owner on that voyage (2) of (8).

Ship Information: TNA, DEL2/108.
Captain Information: Syrett & Charnock (see footnote 41).

References

1. Sir Leoline Jenkins to Mr. Secretary Williamson *Nimeguen* 3 April 1675. William Wynne, *The Life of Sir Leoline Jenkins Judge of the High-Court of Admiralty*, London, 1724. Mathematici are astrologers.

2. Carl E Swanson, 'Privateering in Early America', *Journal of Maritime History*, I (2), 1989, 258-259; John B Hattendorf, *War at Sea in the Middle Ages and the Renaissance*, 1. publ., repr. ed., Warfare in History; Woodbridge [u.a.] Boydell, 2003, 1-24. Historians who agree with Mahan's denigration of privateering include: Julian Stafford Corbett, *Some Principles of Maritime Strategy*, Classics of Sea Power, Annapolis, Md.: Naval Institute Press, 1988, and Herbert W. Richmond, *Statesmen and Sea Power. Based on the Ford Lectures Delivered in the University of Oxford in the Michaelmas Term, 1943*, Westport, Conn., Greenwood Press, 1974.

3. The Oxford English Dictionary lists its first appearance in 1641, it did not enter into common usage until the 1660s. "privateer, *n.*" OED Online. June 2010. Oxford University Press. http://dictionary.oed.com/cgi/entry/50188923. (accessed 31 July 2010).

4. N A M Rodger, 'The military revolution at sea', in N A M Rodger, *Essays in Naval History, from Medieval to Modern*, Variorum Collected Studies Series, Farnham, England ; Burlington, VT, Ashgate, 2009, 240. It is important to note that war was never officially declared between England and Spain during Elizabeth's reign.

5. J R Hill, *The Prizes of War : The Naval Prize System in the Napoleonic Wars, 1793-1815*, Portsmouth, England: Royal Naval Museum Publications: Stroud, Gloucestershire, 1998, 5-8.

6. Still a definite work in field of extraterritorial violence is: Janice E Thomson, *Mercenaries, Pirates, and Sovereigns : State-Building and Extraterritorial Violence in Early Modern Europe*, Princeton Studies in International History and Politics, Princeton, NJ, Princeton University Press, 1994. Janice E Thomson, *Mercenaries, Pirates, and Sovereigns : State-Building and Extraterritorial Violence in Early Modern Europe*, Princeton Studies in International History and Politics, Princeton, NJ, Princeton University Press, 1994. Another excellent article on the same subject from the Dutch point of view is: Louis Sicking, 'State and non-state violence at sea: privateering in the North Sea region since 1550', in Morten Hahn-Pedersen, *Bridging Troubled Waters : Conflict and Co-Operation in the North Sea Region since 1550 : 7th North Sea History Conference*, Dunkirk 2002, 2005.

7. Marco van der Hoeven, *Exercise of Arms : Warfare in the Netherlands, 1568-1648*, History of Warfare, New York, Brill, 1997, 186-189.

8. Grover Clark, 'The English Practice with Regard to Reprisals by Private Persons', *The American Journal of International Law*, 27(4), 1933, 715-722.

9. As both Sicking and Rodger point out, the Dutch had assumed the modern form of privateering earlier than the English with the development of the *kaperbrief* in the late sixteenth century.

10. The National Archives (henceforth TNA), High Court of Admiralty records (henceforth HCA) MSS HCA 25/9.

11. For more on the revolutionary developments pertaining to the High Court of Admiralty in the seventeenth century see: Reginald G Marsden, *A Digest of Cases Relating to Shipping, Admiralty, and Insurance Law, from the Reign of Elizabeth to the End of 1897*, London, Sweet and Maxwell, 1899); Edward Stanley Roscoe, *Studies in the History of the Admiralty and Prize Courts*, London, Stevens & Sons, limited, 1932; D J Ll Davies, *The Development of Prize Law under Sir Leoline Jenkins*, Problems of Peace and War; Variation, Grotius Society.; Transactions, V.21. London, Grotius Society, 1936.

12. Gijs Rommelse, *The Second Anglo-Dutch War (1665-1667) : Raison D'état, Mercantilism and Maritime Strife*, 2006, 124. The National Archives website has a historical currency converter which when given £4000 pounds for the year 1660 gives a result of £307,080 or roughly $478,505 for an approximation of what a bond would cost today. Naturally as the disclaimer on the website states: 'the results of the calculations are intended to be a general guide to historic values, rather than a categorical statement of fact'. http://www.nationalarchives.gov.uk/currency/results.asp#mid

13. The military revolution started out as a hotly debated theory but has become widely accepted; although its outlines are often molded according to each individual author. A few notable works are: Geoffrey Parker, *The Military Revolution : Military Innovation and the Rise of the West, 1500-1800*, Cambridge [England], New York, 1988; Jeremy Black, *A Military Revolution? : Military Change and European Society, 1550-1800*, Atlantic Highlands, N J, Humanities Press, 1991; Brian M Downing, *The Military Revolution and Political Change : Origins of Democracy and Autocracy in Early Modern Europe*, Princeton, N J, Princeton University Press, 1992.

14. A sampling of some of the best recent work on this subject are: John Brewer, *The Sinews of Power : War, Money, and the English State, 1688-1783*, 1st Harvard University pbk. ed, Cambridge, Mass., Harvard University Press, 1990; M. J. Braddick, *State Formation in Early Modern England, C. 1550-1700* (Cambridge [England], New York: Cambridge University Press, 2000; Christopher Storrs, *The Fiscal-Military State in Eighteenth-Century Europe : Essays in Honour of P G M Dickson*, Farnham, England, Burlington, VT, Ashgate, 2009.

15. Discussed throughout: N A M Rodger, 'The military revolution at sea', in N A M Rodger, *Essays in Naval History, from Medieval to Modern*, 59-76. Another excellent investigation of the common interests of commercial and maritime communities of the period is: Natasha Glaisyer, *The Culture of Commerce in England, 1660-1720*, Royal Historical Society Studies in History, New Series; Woodbridge, UK, Rochester, NY, 2006.

16. Peter Lake and Steve Pincus, 'Rethinking the Public Sphere in Early Modern England', *The Journal of British Studies* 45(2), 2006; Paul D Halliday, *Dismembering the Body Politic : Partisan Politics in England's Towns, 1650-1730*, Cambridge Studies in Early Modern British History, New York, 1998.

17. N A M Rodger, 'Queen Elizabeth and the myth of sea-power in English history', in N A M Rodger, *Essays in Naval History, from Medieval to Modern*, 153-174.

18. For more on how absolutist states evolved differently see: David Parrott, *Richelieu's Army : War, Government, and Society in France, 1624-1642*, 'Digitally printed 1st pbk.' ed., Cambridge Studies in Early Modern History, Cambridge, New York, Cambridge University Press, 2006; William Beik, *Absolutism and Society in Seventeenth-Century France : State Power and Provincial Aristocracy in Languedoc*, Cambridge Studies in Early Modern History, Cambridge Cambridgeshire, New York, Cambridge University Press, 1985; Jan Glete, *Absolutism or Dynamic Leadership? The Rise of Large Armed Forces and the Problem of Political Interest Aggregation from a Mid-17th Century Perspective*, Suecoromana, 4, (Stockholm: [Svenska institutet i Rom], 1997.

19. All of the surviving records of that kind for the Second Anglo-Dutch Naval War are contained in TNA, HCA 25/9 and HCA 25/10. All of the surviving records for the Third Anglo-Dutch Naval War are contained in the TNA, DEL 2/108. In addition, TNA holds voluminous records pertaining to other aspects of privateering such as adjudication; the relevant prize court papers and prize court judgments and appeals among others. Major archives that provided sources for the entirety of this project include The National Archives, The British Library, the National Maritime Museum, the Bodleian Library, the Pepys Library at Magdalene College, the Folger Shakespeare Library, and the Library of Congress.

20. Warrants typically contain the privateer ship's name, the captain's name, tonnage of the ship and the date the warrant was signed, by James, Duke of York.

21. The numbers assigned in the register book appear also on the warrants, bonds and other materials themselves to serve as a universal identification number assigned by the state to better keep track and regulate the actions of privateering enterprise building upon the record keeping lessons and experiences of the previous war.

22. In the *Treaty of Marine between Great Britain and Holland London February 17, 1668* [British Library MSS Additional Manuscripts 61913- Treaties 1667-1758] the number of crewmen now dictates the amount of the bond paid to the state before the privateer could set sail; a privateer ship with over 150 crew would have to pay a £3000 bond and those under 150 crew would have to pay £1500 bond. Hence the need for ever greater record keeping and scrutiny by the state in its regulation of privateering enterprise. This appears to be a major shift in policy after the Second Anglo-Dutch War in which all privateer ships were required to pay a £4000 bond regardless of the size of the crew.

23. It is remarkable how well the international system worked in European waters in regards to privateering enterprise by the second half of the seventeenth century; popular misconceptions about privateering being synonymous with piracy stem from the legendary depredations of privateers in the Americas who operated outside the intricate regulation system and respected precedents of the 'Law of the Seas' that successfully curbed piracy in European waters.

24. Eric J Graham, *A Maritime History of Scotland, 1650-1790*, East Linton, East Lothian Scotland, Tuckwell, 2002; Eric Graham, 'Scottish Marine During the Dutch Wars', *Scottish Historical Review* Vol. LXI, no. 1, 1981. Although Welsh privateers were included in my database, an excellent localized study of Welsh privateering is: J D Davies, 'The Revenge of Llanelli: A Welsh Privateer of the Seventeenth Century', *Maritime Wales*, Vol. 21, 2000. See later footnotes for Nuala Zahedieh's significant scholarly analysis of privateering in the Americas; no such work has been done yet for privateers based from Tangier, an issue the author of this paper is researching.

25. Privateering was finally abolished by all the major European powers in the 1856 Declaration of Paris.

26. N A M Rodger, 'The new Atlantic: naval warfare in the sixteenth century', in N A M Rodger, *Essays in Naval History, from Medieval to Modern*, 234-247.

27. It is no coincidence that English colonization of the Americas starts in earnest after the cessation of hostilities in 1604.

28. A prominent example of these early charter companies is the Company of Merchant Adventurers of London. The authoritative work in this field remains: Robert Brenner, *Merchants and Revolution : Commercial Change, Political Conflict, and London's Overseas Traders, 1550-1653*, Princeton, N J, Princeton University Press, 1993.

29. Kenneth Gordon Davies, *The Royal African Company*, London, New York, Longmans, 1957, 24-31.

30. The colony just barely survived chronic mismanagement since its founding in 1607 leading up to a concerted attack by Native Americans that almost wiped that colony out in 1622; hence the decision by the Crown to assert its control and revoke the charter.

31. Still an authoritative work on the evolution of maritime law in general is: Reginald G Marsden, *Documents Relating to Law and Custom of the Sea*, 2 vols, Union, N J, Lawbook Exchange, 1999. Several recent analyses of Grotius' primacy in the development of the maritime law are: Hugo Grotius and Martine Julia Van Ittersum, *Commentary on the Law of Prize and Booty*, Natural Law and Enlightenment Classics, Indianapolis, Liberty Fund, 2006); Hedley Bull, *Hugo Grotius and International Relations*, 1992.

32. Davies, *The Royal African Company*, 22-23.

33. For more on these 'Letter of Marque ships' as they became known, which became more prevalent in the eighteenth century, see: David J Starkey, *British Privateering Enterprise in the Eighteenth Century*, Exeter Maritime Studies , No. 4, Exeter, Devon, University of Exeter Press, 1990, 49-52.

34. Richard Lawrence Ollard, *Man of War: Sir Robert Holmes and the Restoration Navy*, London, Hodder & Stoughton, 1969, 63.

35. Elizabeth Donnan and Carnegie Institution of Washington. Division of Historical Research., *Documents Illustrative of the History of the Slave Trade to America*, 4 vols., Carnegie Institution of Washington. Publication, Washington, DC, Carnegie Institution of Washington, 1930, 169-172.; Royal African Company. 1667. *The several declarations of the Company of Royal Adventurers of England Trading into Africa inviting all His Majesties native subjects in general to subscribe and become sharers in their joynt-stock ... : as also a list of the Royal Adventurers of England Trading into Africa*. England: s.n.

36. Charles McLean Andrews, *British Committees, Commissions, and Councils of Trade and Plantations, 1622-1675*, Johns Hopkins University Studies in Historical and Political Science, Ser. 26, No. 1-3, Baltimore, The Johns Hopkins Press, 1908, 68. as quoted in Donnan and Carnegie Institution of Washington. Division of Historical Research, *Documents Illustrative of the History of the Slave Trade to America*., 170.

37. Glaisyer, *The Culture of Commerce in England, 1660-1720*. states that these middling sort merchants, who are incidentally the backbone of privateering enterprise, are by the end of the seventeenth century major investors in government securities and thus play a fundamental role in the 'financial revolution' that makes England a model fiscal-military state.

38. For a detailed account of this infamous expedition see: Ollard, *Man of War: Sir Robert Holmes and the Restoration Navy*, 85-119. It is also important to note that an equally provocative expedition successfully captured New Netherland in North America from the Dutch in August of 1664 as well.

39. C S Knighton reveals that a parliamentary committee created to investigate the 'abuses' inflicted on CRAETA by the Dutch off the coast of Africa during 1663-4 was packed with nine members of CRAETA including Sir Richard Ford, a man prominent in privateering enterprise. C S Knighton, *Pepys and the Navy*, Stroud, Gloucestershire, Sutton Pub., 2003, 54-55.

40. N A M Rodger, *The Command of the Ocean : A Naval History of Britain, 1649-1815*, New York: W W Norton, 2005, 68.; Ollard, *Man of War: Sir Robert Holmes and the Restoration Navy*, 129-130.

41. TNA, HCA 25/9, HCA 25/10.

42. Ibid.; G E Aylmer, 'Noell, Sir Martin (bap. 1614, d. 1665)', P R S Baker in *Oxford Dictionary of National Biography*, ed. H C G Matthew and Brian Harrison, Oxford, OUP, 2004; online ed., ed. Lawrence Goldman. http://www.oxforddnb.com/view/article/37814 (accessed 31 July 2010).

43. David Syrett and R L DiNardo, *The Commissioned Sea Officers of the Royal Navy, 1660-1815*, Occasional Publications of the Navy Records Society, Vol. 1, Aldershot, England Brookfield, VT, USA, Scolar Press for the Navy Records Society, Ashgate Pub. Co., 1994.

44. I have no record of this 1670-71 consortium from TNA sources but it is mentioned in: J K Laughton, 'Modyford, Sir James, baronet (1618–1673)', rev. Nuala Zahedieh, in *Oxford Dictionary of National Biography*, ed. H C G Matthew and Brian Harrison, Oxford, OUP, 2004; online ed., ed. Lawrence Goldman, January 2008, http://www.oxforddnb.com/view/article/18870 (accessed 31 July 2010); Zahedieh, in a masterful overview of privateering enterprise based in Jamaica, notes that the 'gentlemen at court' who owned the *Lilly* saw little profit from their consortium. Nuala Zahedieh, 'A Frugal, Prudential and Hopeful Trade. Privateering in Jamaica, 1655–89', *The Journal of Imperial and Commonwealth History*, 18(2), 1990, 156.

45. Henry Morgan and 'privateers' operating in the Americas (and unscrupulous governors such as Modyford who authorized and dispatched them) paid little heed to the strict rules and regulations that governed privateering in Europe; this was not considered unusual since the conduct of warfare in the Americas was often quite removed from the accepted norms of warfare in Europe - distance, lack of authority and the concept of 'no peace beyond the line' meant that warfare in the Americas was less 'civilized' and far less subject to regulation than in Europe.

46. For a classic perspective on the demise of CRAETA that differs from Zook's interpretation (reference 52) see: Henry Meredith, *An Account of the Gold Coast of Africa, with a Brief History of the African Company*, Cass Library of African Studies, Travels and Narratives, No. 20, London, Cass, 1967.

47. TNA, DEL 2/108; J K Laughton, 'Modyford, Sir James, baronet (1618–1673)', rev. Nuala Zahedieh, in *Oxford Dictionary of National Biography*, ed. H C G Matthew and Brian Harrison, Oxford:, OUP, 2004; online ed., ed. Lawrence Goldman, January 2008, http://www.oxforddnb.com/view/article/18870 (accessed 31 July 2010); J D Davies, 'Kempthorne, Sir John (c.1620–1679)', in *Oxford Dictionary of National Biography*, ed. H C G Matthew and Brian Harrison, Oxford, OUP, 2004; online ed., ed. Lawrence Goldman, January 2008, http://www.oxforddnb.com/view/article/15340 (accessed 31 July 2010).

48. This information coming from the few affidavits contained in TNA 25/9; the actual number of affidavits with connections to the Africa House could be much higher had they survived. Later the RAC moves to Leadenhall Street, London.

49. TNA, HCA 25/9, HCA 25/10. An excellent detailed study of warship design and construction in general during this period can be found in: Richard Endsor, *The Restoration Warship : The Design, Construction and Career of a Third Rate of Charles II's Navy*, Annapolis, Md., Naval Institute Press, 2009.

50. D W Hayton, 'Leighton, Sir Elisha (d. 1685), In *Oxford Dictionary of National Biography*, edited by H C G Matthew and Brian Harrison, Oxford: OUP, 2004. Online ed., edited by Lawrence Goldman, May 2006. http://www.oxforddnb.com/view/article/16398 (accessed 31 July 2010); National Maritime Museum MSS CLI 37, 'Instructions for Commissioners to Ellis Leighton'.

51. An interesting comparison should be noted here. Zahedieh notes that the usual payment for a bond in Jamaica during the Restoration was at its highest half this amount and could range anywhere from a scant £200 to £2000. This is certainly an indication of the scope and power of the state's regulation of privateering enterprise in England in comparison to Jamaica and demonstrates just how little toleration there was for any piratical or openly brazen acts of embezzlement in European waters. Zahedieh, 'A Frugal, Prudential and Hopeful Trade. Privateering in Jamaica, 1655–89',", 149.

52. Matching data from TNA, HCA 25/9, TNA, HCA/10 to: Syrett and DiNardo, *The Commissioned Sea Officers of the Royal Navy, 1660-1815* and John Charnock, *Biographia Navalis; or, Impartial Memoirs of the Lives and Characters of Officers of the Navy of Great Britain, from the Year 1660 to the Present Time; Drawn from the Most Authentic Sources, and Disposed in a Chronological Arrangement*, London, R. Faulder, 1794. An excellent source for finding further research on a particular officer in the seventeenth century is: Peter Le Fevre, 'Re-Creating a Seventeenth-Century Sea Officer', *Journal for Maritime Research*, 2001.

53. Charnock, *Biographia Navalis*, Vol. 1, 81.

54. Charnock, *Biographia Navalis*, Vol. 1, 84.

55. TNA, PROB 11/324.

56. George Frederick Zook, *The Company of Royal Adventurers Trading into Africa*, Lancaster, PA., Press of the New Era Printing Co., 1919, 23-27.

57. Charles II - volume 180: December 1-7, 1666, *Calendar of State Papers Domestic: Charles II, 1666-7*, 1864, 307-328.

58. *London Gazette*, Issue 224, 9 January 1666[7], 1.

59. The captain's occupation and status is listed if they are an owner on the bond; hence the figure of 87 identifiable captains out of 100 voyages for the Second Anglo Dutch War, TNA, HCA 25/9; 24 identifiable captains out of 30 voyages for the Third Anglo Dutch War, TNA, DEL2/108.

60. TNA, HCA 25/9; Syrett and DiNardo, *The Commissioned Sea Officers of the Royal Navy, 1660-1815*.

61. An excellent in depth treatment of this topic is: J D Davies, *Gentlemen and Tarpaulins : The Officers and Men of the Restoration Navy*, Oxford Historical Monographs, Oxford New York, Clarendon Press, Oxford University Press, 1992.

62. N A M Rodger, 'Training or education: a naval dilemma over three centuries', in Rodger, *Essays in Naval History, from Medieval*

	to Modern, XVII-XVII.
63	J D Davies, 'Holles, Sir Frescheville (1642–1672)', *Oxford Dictionary of National Biography*, Oxford University Press, September 2004; online edn, Jan 2008 [http://www.oxforddnb.com/view/article/13551] (accessed 31 July 2010).
64	J D Davies, *Pepys's Navy : Ships, Men & Warfare, 1649-1689*, 2008, 94-96.
65	This is a stark contrast to charter companies such as CRAETA; The 'middling sort' only comprise 25% of the 1660 CRAETA charter and 32% of 1667 CRAETA charter; data taken from charters analyzed in identical sources mentioned in footnote 27.
66	TNA, HCA 25/9; Charnock, *Biogrpahia Navalis*, Vol. 1, 294.
67	Listed in appendixes from: Frank L Fox, *A Distant Storm : The Four Days' Battle of 1666 : The Greatest Sea Fight of the Age of Sail*, Rotherfield, Press of Sail Publications, 1996. For more on how merchantmen were still enlisted as late as the Restoration period to fight alongside state owned purpose built warships see: Frank L Fox, 'Hired Men-of-War, 1664-67 part I', *The Mariner's Mirror*, Vol. 84, London, 1998., 13-25; 'Hired Men-of-War, 1664-67 part II', *The Mariner's Mirror*, Vol. 84, London, 1998, 152-172.
68	Davies, *Gentlemen and Tarpaulins : The Officers and Men of the Restoration Navy*, 80.
69	TNA, HCA 30/821 *Warrants for Appointments January 12 1666 Conventry to Duke of York*. As mentioned previously 'letter of marque' privateers were not purpose-designed warships intended solely for cruising for prizes but merchantmen and part-time opportunistic maritime predators.
70	J R Powell et al., *The Rupert and Monck Letter Book. 1666: Together With Supporting Documents*, Publications of the Navy Records Society, V. 112, London, Printed for the Navy records Society, 1969, 78.
71	C R Boxer, 'Some Second Thoughts on the Third Anglo-Dutch War, 1672-1674', *Transactions of the Royal Historical Society*, 19, Royal Historical Society, 1969, 75. It is also possible that records of additional voyages were lost and thus not contained in TNA collections.
72	This information comes from the bonds as the captain's name would be listed as an owner on the bottom of the bond.
73	TNA, HCA25-9/10; DEL2-108; Syrett and DiNardo, *The Commissioned Sea Officers of the Royal Navy, 1660-1815*.
74	*ibid*.
75	*ibid*.
76	Two suggested general biographies of Pepys, one a classic one recent: Richard Lawrence Ollard, *Pepys : A Biography*, London: Hodder and Stoughton, 1974; & Claire Tomalin, *Samuel Pepys : The Unequalled Self*, London: Viking, 2002.
77	Knighton, *Pepys and the Navy*, 10.
78	Mynors Bright et al., *The Diary of Samuel Pepys*, London, New York, G. Bell and sons, 1893. Project Gutenberg, June, 2003, [EBook #4200], http://www.gutenberg.org/dirs/4/2/0/4200/4200.txt., henceforth 'Diary,' 26 September 1666.
79	Charles II - volume 168: 17-25 August 1666, *Calendar of State Papers Domestic: Charles II, 1666-7*, 1864, 47-67. We also learn from a dispatch on 15 December 1666 from Silas Taylor that 'The *Flying Hound* was built by the Shoemakers' Company of Amsterdam.' Charles II - volume 181: 8-15 December 1666, *Calendar of State Papers Domestic: Charles II, 1666-7*, 1864, 328-350.
80	TNA, HCA 25/9.
81	Diary, 23 October 1666.
82	Starkey, *British Privateering Enterprise in the Eighteenth Century*, 73-77.
83	Information on the activities of the *Flying Greyhound* during this period are found in Pepys' diary, Calendar of State Papers Domestic, the *London Gazette* and significantly a collection from the Pepys Library at Magdalene College (MSS 2871 665, 667) entitled *Papers to my Account about the Flying Greyhound and the St John Baptist Prize & the Rose Prize*.
84	Ollard, *Pepys : A Biography*, 202-203.
85	Diary, 22 December 1666.
86	Diary, 31 December 1666.
87	Sweden's close ties to England stem from both a historically close relationship recently brought to the forefront by Cromwell's diplomacy during the Interregnum, but also lie in long term mutual geo-political aims shared by both states in their desire to stem the designs of both the Dutch and their allies the Danes in the Baltic.
88	*Orders in Council 12th of September 1666 at the Court of Whitehall*, TNA, PC 2/59, pg 161.
89	Jenkins revolutionized the High Court of Admiralty and defended and further defined its jurisdiction, see: Davies, *The Development of Prize Law under Sir Leoline Jenkins*.; D F Taylor, 'Sir Leoline Jenkins, 1625-1685', PhD Diss., University of London, 1973; Wynne, *The Life of Sir Leoline Jenkins Judge of the High-Court of Admirality*, London, 1724.
90	An anonymous tract in the British Library lists all the tricks and trades of these forgeries: MS Additional 34729 West Papers FF 180. Two excellent articles dealing with the pitfalls of adjudication and neutrality issues (with specific reference to Sweden) are: Leos Müller, *Swedish Shipping Industry: A European and Global Perspective, 1600-1800*, vol. 6, Journal of History for the Public, Department of Occidental History, Osaka University, 2009; Steve & Little Murdoch, Andrew & Forte, A D M, 'Scottish Privateers, Swedish Nuetrality and the Third Anglo-Dutch War', *Forum Navale*; Sjöhistoriska Samfundet, 59, 2003.
91	TNA, SP 9/240, 17-18.
92	Davies, *Pepys's Navy : Ships, Men & Warfare, 1649-1689*, 104; Rommelse, *The Second Anglo-Dutch War (1665-1667) : Raison D'état, Mercantilism and Maritime Strife*, 124-126.
93	TNA, SP 9/240, 19-20.
94	Diary, 21 January 1666[7].
95	Diary, 21 March 1666[7].
96	An in depth examination of the scandal can be found in: Richard Lawrence Ollard, *Cromwell's Earl : A Life of Edward Mountagu 1st Earl of Sandwich*, London, Harper Collins, 1994.
97	'Charles II - volume 196: April 1-9, 1667', *Calendar of State Papers Domestic: Charles II, 1667*, 1866, 1-23.
98	Pepys Library at Magdalene College (MSS 2871 665, 667) entitled '*Papers to my Account about the Flying Greyhound and the St John Baptist Prize & the Rose Prize*'.
99	Diary, 25 July 1667.
100	Knighton, *Pepys and the Navy*, 80-81.
101	Diary, 1 March 1667[8].
102	Diary, 20 April 1668.
103	*London Gazette*, Issue 652; 12 February 1671.

Richard Brabander *is a PhD candidate in the History Department at Brandeis University in Waltham, USA. His dissertation is entitled* A Private Interest in Publick Warr; an Analysis of English Privateering 1660-1685.

CHATHAM TO ERITH VIA DOVER. CHARLES II's SECRET FOREIGN POLICY AND THE PROJECT FOR NEW ROYAL DOCKYARDS, 1667-1672*

J D Davies

Abstract

The secret Treaty of Dover, 1670, remains one of the most controversial events of Charles II's reign. This paper seeks to explain one of its most important but neglected terms, the proposed English territorial acquisitions in the Netherlands, in the context of aspects of naval policy, notably the little-known project to create an entirely new royal dockyard at Erith.

Introduction. The Dutch Medway Raid

At about 10 a.m. on the morning of Wednesday 12 June 1667, a squadron of Dutch warships sailed up Gillingham Reach on the River Medway. Ahead of them lay a large chain, stretched taut across the river, blocking their way to the British warships that lay beyond, off the great naval dockyard at Chatham.[1] Most of the British ships were dismasted and virtually unarmed. Lacking the money to send a proper fleet to sea for that summer's campaign (and believing in any case that peace was imminent), King Charles II had ordered the ships to be laid up, trusting that the chain and the forts guarding the Medway would be sufficient to protect the navy against just such a Dutch attack. But most of the forts were still incomplete, and the largest and most important of them, that at Sheerness, had already fallen to the Dutch two days earlier. Still, the great chain appeared to be an insuperable obstacle, and so it might have proved but for the audacity of Jan Van Brakel, a Rotterdam captain, who volunteered to lead his ship, the *Vrede*, in an attack on the barrier. Under heavy fire, he attacked the guardship *Unity*, which protected the chain, and thanks to a supine defence by her inadequate crew, he took her without a serious fight. This allowed the fireship *Pro Patria* to sail directly at the chain, which broke on impact (according to the Dutch) or else sank under its own weight (according to the English).[2] Beyond one last and easily negotiated barrier of undermanned guardships lay the most seaward of Charles II's great ships, the *Royal Charles*. Only 32 of her 82 guns were still aboard, and she had virtually no crew embarked. The men ordered in haste to tow her to safety up river simply turned and fled when they saw that they were too few, and too weakly armed, to resist the approaching Dutch. A small prize crew quickly took possession of the ship, striking her British colours and replacing them with the tricolour of the United Provinces of the Netherlands.

The Dutch resumed their attack on the following day, albeit under heavy fire from Upnor Castle and other batteries that had been hastily erected on both sides of the Medway. Three more great ships lying scuttled in shallow water – the *Royal James*, *Royal Oak* and *Loyal London* - were attacked and set on fire. Friday 14 June brought further humiliation. Despite the immense navigational difficulties presented by the River Medway, and under fire from British forces ashore, the Dutch managed to get the *Royal Charles* down river to Sheerness, where the Dutch flag still flew over the fort, and thence back to Holland. The main Dutch fleet, commanded by Michiel De Ruyter, remained in the mouth of the Thames, and on 2 July troops from it made a substantial landing near Felixstowe in order to attack Landguard Fort, which protected the small dockyard at Harwich on the opposite shore. The defences and defenders of Landguard proved more resilient than those of Sheerness fort and the *Royal Charles*, and this second Dutch invasion of English soil within three weeks was beaten off. However, a Dutch squadron remained in the mouth of the Thames until the middle of July, while De

---

* *Paper presented at the fourteenth annual conference of the Naval Dockyards Society held at the National Maritime Museum, Greenwich, on 17 April 2010. Theme: Pepys and Chips, Dockyards, Naval Administration and Warfare in the Seventeenth Century.*

Ruyter's main fleet cruised unchallenged along the south coast of England until after peace between Britain and the Netherlands was finally concluded at Breda on 21 July.

Fig. 1. The Dutch attack on the Medway, 1667, showing Sheerness fort under occupation (right) and the river Medway up to Rochester bridge (top right). Trustees of the British Museum.

The Aftermath of the Medway Raid

In material terms, the Medway debacle could have been much worse. The Dutch could have pressed on and destroyed the other scuttled ships, the dockyard itself, and the important commercial timber yards nearby.[3] As it was, the attack caused untold damage to the reputations of both king and country; or at least, Charles II clearly believed that it did. The Dutch onslaught was regarded, quite categorically, as an invasion.[4] Seventeenth century commentators did not distinguish between invasions of outright conquest and mere incursions, regarding both as a dire affront to a nation's dignity and a humiliating demonstration of national weakness.[5] Many of Charles's actions in the period 1667-1672 can be explained by his determination to obliterate the damage done to his and his country's reputation by the humiliating Dutch invasion, a determination that focused upon two 'r's: reputation and revenge. Only a month after the disaster, one of his ministers reported that 'His Majesty resolves though we have peace to have a strong fleet next year [as] it will be but what is necessary after the affronts we have received to show we resolve still to be masters of the sea'.[6] Charles made good his resolution. Some fifty large ships, including one first rate (the new *Charles*, launched at Deptford on 3 March 1668[7]), six seconds and up to 17 thirds, were ordered to sea in the summer of 1668, but it is not clear how many of these actually sailed before the crippling cost led to the premature paying off of the fleet.[8] Even so, the ships that did put to sea were far larger than a normal peacetime Summer Guard, and were clearly intended to send a powerful signal to all of Charles II's potential allies and enemies. So, too, was the name given to the infant prince born on 14 September 1667 to the Duchess of York and her husband, Charles's brother, heir presumptive and Lord High Admiral. The new baby at once became second in line to the throne, yet he was christened Edgar, a name previously unknown in the Stuart dynasty (and in every other dynasty since the Norman conquest). The choice mystified some, but others recognised the symbolism: Edgar, a tenth century King of England, was traditionally regarded as the founder of English sea power, and the king who first asserted sovereignty over the seas around his kingdom.[9] The choice of such an apparently eccentric name for their potential successor could only mean that Charles and James intended the future King Edgar to exercise an equally unchallenged sovereignty at sea; and at some point, the attainment of that goal would inevitably entail the obliteration of the humiliating damage done to the country's naval reputation by the attack on Chatham. For that, the Stuart brothers would need ships, and the ships would need dockyards.

Before 1667, the English state had built just four three-decked, first rate men-of-war. During the six years from 1667 to 1673 it completed the astonishing total of six such vessels,[10] two at Portsmouth, two at Deptford and one each at Chatham and Woolwich, along with a large second rate at Portsmouth that was re-rated as a first within three years of its launching. Of course, it cannot be proved that Charles II built these ships with the explicit intention of avenging Chatham by literally blasting the Dutch navy off the oceans. Almost all of them were direct replacements for great ships lost during the previous war, and several had been ordered well before the Medway raid took place. In most cases, too, the new 'great ships' bore the same names as the war losses that they replaced. But this in itself could be evidence that Charles sought to send a clear signal to the Dutch that his navy could rapidly make good its losses by building a new and awesomely powerful generation of men-of-war bearing the same names as those they were intended to avenge.[11]

The French Threat and Espionage in the French Dockyards

The Dutch were not the only potential opponents for the new leviathans that emerged from the royal dockyards in the years that followed the Medway raid. In 1661 the French navy consisted of 31 ships, only one of which was larger than a third rate. Ten years later it comprised 123 ships, over a fifth of which were first and second rates. In 1668-1669 alone, six first rates were launched at Toulon and Brest; the largest, the *Royal Louis* and *Soleil Royal*, could carry up to 120 guns each, twenty more than the greatest of Charles II's ships, and were the largest ships in the world. This phenomenal expansion did not go unnoticed on the other side of the Channel. Between 1668 and 1671, government ministers received countless reports of Louis XIV's building programme from merchantmen trading with French ports and other sources of intelligence.[12] Charles II was certainly not oblivious to this remarkable French construction programme. Even if he retained his confidence in Louis XIV's friendship, and even if he convinced himself that the new French navy was intended only to further Louis's ambitions against the Dutch or the Spanish and thus presented no threat to the British kingdoms,[13] its existence was not something that Charles could simply ignore. Consequently, in the late summer of 1668 Charles and Arlington despatched a spy, a Portsmouth shipwright named Thomas Castle, to reconnoitre the French Atlantic dockyards.[14] Castle was at La Rochelle by 22 September, and gathered intelligence both on the ships in the yard and the forthcoming French expedition to St Christopher in the Caribbean. He remained at La Rochelle for over a fortnight, reporting the alarming news that the French intended to have over a hundred ships at sea in the summer of 1669, and then moved on to Brest, reaching the Breton yard some time in the middle of October and staying for over a month.[15] Castle produced a somewhat sketchy plan of the harbour (although he managed somehow to take his own soundings) and provided detailed reports both on the expansion of the yard, especially the design of the new wet dock and ropery, and on the ships under construction there.[16] One ship in particular attracted his attention: this was 140 ft long in the keel and reported to be intended for 120 guns, but Castle counted only 16 gunports on the lower tier, 15 on the middle and 13 on the upper, so he reckoned she would be unlikely to mount more than 110. Even so, this would have made the new ship – almost certainly the *Royal Duc*, renamed *La Reine* in 1671[17] – more heavily armed than anything that Charles II possessed, and Castle reported that she was very high between decks, '[so] that I cannot near reach the upper edge of the beam with my hands and stand on my foots *[sic]*'.[18] Castle also reported the construction of 'a new yard made at a place called Rochford [Rochefort]', talked freely with ships' captains, mast makers and other dockyard workers, and sent back a number of detailed lists of the French navy.[19] During November the recipient of Castle's letters, Arlington's deputy Joseph Williamson, drew up several memoranda based on these and other intelligence, noting in particular the impressive speed and scale of the works at Rochefort, as well as the efforts that were being made to ready several squadrons for sea.[20]

Castle's mission and reports probably contributed to the significant change of

Fig. 2. The *corderie royale* at Rochefort, the new French royal dockyard being completed in 1668-9. Author's photograph.

attitude towards France and her maritime pretensions that began to appear from the autumn of 1668 onwards in the king's correspondence with his sister Henrietta, known as 'Madame', the wife of Louis' deranged, cross-dressing brother Philippe, the Duc d'Orleans (or 'Monsieur'). Even before Castle's first despatch arrived from La Rochelle, Charles informed 'Madame' that

the first [obstacle to an Anglo-French treaty] is the great application there is at this time in France to establish trade and to be very considerable at sea, which is so jealous a point to us here who can be only considerable by our trade and power at sea, as any steps that France makes that way must continue a jealousy between the two nations which will upon all occasions be a great hindrance to an entire friendship, and you cannot choose but believe that it must be dangerous to me at home to make an entire league till first the great and principal interest of this nation be secured whic is trade.[21]

The king returned to this theme in January 1669, when he informed Henrietta that the only obstacle to an alliance with Louis was 'the matter of the sea', a point so important to Charles that he could not be 'answerable to my kingdoms if I should enter into an alliance wherein their present and future security were not fully provided for'.[22] By the summer, Charles was demanding that Louis should suspend all naval construction as a precondition of an alliance against the Dutch. Louis was initially furious, but he eventually agreed to a one-year moratorium.[23] However, the 'Sun King' could afford to be magnanimous; his building programme of large ships was effectively already complete, so his 'concession' could hardly have been emptier.

The Diplomatic Context

The diplomacy of the years 1667-1672 was an intricate confection of duplicity, tragedy, brinkmanship and high farce. At the root of much of it was the precarious life of one child, a child born with appalling physical and mental disabilities. On 7 September 1665, some weeks short of his fourth birthday, the child had become Carlos II, King of Spain. The product of generations of inbreeding among the Habsburg dynasty, the accession of Carlos *el Hechizado* – 'the bewitched' – immediately made the question of the succession to the Spanish throne the dominant issue in European politics. The Spanish empire was no longer the invincible global power it had once been. Even so, Spain still ruled over much of southern Italy, all of Flanders and most of south and central America, all potential colonies (and markets) for whoever could make good their claim to succeed Carlos. Louis XIV was married to *el Hechizado*'s elder sister, Maria Theresa, who had renounced her rights to the Spanish throne upon marriage; but the renunciation had been conditional on the payment of 500,000 escudos, and Spain never had any realistic chance of raising that sum. More immediately, Louis claimed that certain laws in parts of the Spanish Netherlands gave the children of a first marriage (even if female) a superior claim to property than any children of a second (even if male), and he used this tenuous legal claim of 'devolution' to justify a full-scale invasion of Flanders, aiming to annex large tracts of territory.

The main French field army advanced into the Spanish Netherlands in May 1667.[24] A

Fig. 3. Henrietta Anne, Duchesse d'Orléans, the youngest sibling of King Charles II, and his intermediary with her brother-in-law Louis XIV. Trustees of the British Museum.

string of important towns fell quickly, taking French-held territory to within some 20 miles of the Dutch border.[25] The speed and ease of French progress thoroughly alarmed the surrounding states. The Dutch were still notionally allied with France in their war against Britain, and their fleet was causing havoc in the Thames and Medway at the exact same moment that Louis' armies were advancing unstoppably down the valleys of the Meuse and Dandre. The need to free themselves to face the potential threat from France contributed to the speed with which the Dutch agreed to terms with Charles II's negotiators in the Treaty of Breda, a treaty which was remarkably generous to the country that had been humiliated at Chatham so recently and comprehensively. Lord Arlington, Charles II's pro-Spanish secretary of state, then embarked on a series of intricate negotiations with both the French and the Dutch, hoping to prise them apart. Arlington favoured an alliance with the Dutch, but Charles II favoured the opposite course. His personal inclinations were invariably francophile (though not uncritically so), while his great friend (and Arlington's bitter rival) the Duke of Buckingham took a similar stance, albeit one born of blatant political opportunism. In December 1667, therefore, a proposal for an Anglo-French alliance against the Dutch was put to Louis' negotiators and rejected out of hand; limited and insincere French counter-proposals were received unfavourably. Oblivious to Dutch alarm at the progress of his 'war of devolution', Louis still believed that his Dutch alliance was more important to his ambitions in the Spanish Netherlands than any new agreement with beaten and bankrupt Britain, and Charles and his ministers were desperate for any diplomatic initiative that would restore something of their shattered reputation.[26]

The failure of the negotiations between Charles and the French ensured the rapid agreement of the 'Triple Alliance', concluded between Britain, Sweden and the United Provinces during January 1668. This contained a secret clause committing the new allies to joint action against the French if they rejected the concessions that Spain was expected to make. Charles quickly leaked this clause to Louis, who was furious at what he regarded as a betrayal by his long-time allies, the Dutch; allies on whose behalf he had gone to war against Britain, and who were now repaying him with such gross ingratitude. Charles's duplicity in revealing the secret clause thus had its desired effect: a wedge was driven between the French and Dutch.[27] Nevertheless, Charles II seemed at first to hold true to the principles of the Triple Alliance, despite some minor disagreements with his new Dutch allies, and the king may even have seen a potential role for himself as the 'arbiter of Christendom', enforcing a balance of power between the western European states.[28] Moreover, thoughts of revenge for Chatham seemed to have been displaced, at least temporarily, by growing concern over French naval construction. Over the winter of 1668-1669 Charles put together a proposal for an offensive and defensive alliance with France, initially floating the suggestion in his private correspondence with his sister Henrietta.[29] The alliance was to be conditional on the suspension of the French building programme and the continuation of the Triple Alliance; there was no mention as yet of Charles's kingdoms going to war against the Dutch. The most controversial aspect of the proposed alliance was Charles's expressed intention to declare himself a Catholic, a potentially explosive undertaking that was shared only with a very small and trusted inner circle of advisors. It is hard to argue with the suggestions that although Charles may have experienced a genuine but short-lived burst of enthusiasm for joining the Church of Rome,[30] the putative conversion was primarily bait that Louis XIV would find irresistible. Moreover, the subsidy that the French king promised to pay in return for Charles making the declaration would be a useful addition to the desperately weak royal finances. But Louis wanted more for his money, insisting that the alliance should culminate in a joint attack on the United Provinces.

At first there was seemingly little enthusiasm for the attack on the English side, and only desultory negotiations for a purely commercial treaty took place.[31] By December 1669, though, Charles had accepted the logic of war with the Dutch, and made a grandiose series of demands on Louis: £1 million before the war began, £600,000 a year for its duration, a chunk of Dutch territory, and, if and when Louis made good his claim to the Spanish succession, Ostend, Minorca and a large part of Spain's overseas empire.[32] Horse trading over the precise terms took place during the months that followed, and the agreement was finally sealed by the apparently innocent device of a family reunion. Princess Henrietta arrived at Dover on 17 May and stayed for two weeks, during which she was feted relentlessly by her two brothers. Henrietta eventually left Dover on 2 June, and Charles and James returned to Whitehall on the following day.[33] The most significant outcome of the visit was the secret treaty signed on 22 May. Its terms were relatively short and straightforward by the standards of international diplomacy, but they were breathtakingly grand in scope. Charles was to make a public profession of the Catholic faith, and would then receive from Louis the sum of two million crowns within six months; but the date of this declaration was left entirely to Charles's discretion. If 'new rights to the Spanish monarchy' came to the King of France – a none-too-subtle euphemism for the constantly expected death of Carlos - then Charles II would aid him in securing and maintaining those rights. The two Kings were to declare war against the United Provinces; Louis was to attack them by land, assisted by 6,000 English troops, while Charles would set out 50 men-of-war to join with 30 French ships, the whole to be under the command of the Duke of York. Charles would receive Walcheren, the mouth of the Scheldt, and the isle of Cadzand as his share of the conquered provinces.[34]

The Treaty of Dover: the River Scheldt and the Dutch Ports

The centrepiece of the Treaty of Dover appears to be the 'catholicising clause', the shocking statement of King Charles II's apparent intent to return his country to Rome, funded by French subsidies. Even in an allegedly secular and multicultural age, a nation that has only just got round to permitting its royal heirs to marry Catholics still has difficulty coming to terms with what seems to be the dastardly betrayal of 'Englishness' implicit in the Treaty of Dover. Much less attention has been paid to the second half of the treaty, yet arguably, from Charles's viewpoint this was more substantive and significantly more important than the very vague commitment to declare himself a Catholic at some indeterminate point in the future. Most historians agree that even if Charles had briefly experienced some sort of genuine enthusiasm for his impending conversion, this had evaporated long before the Treaty of Dover was signed. However, a rather more specific (and, by implication, time-limited) outcome of the expected destruction of the Dutch state would be the acquisition of 'Walcheren, the mouth of the Scheldt, and the isle of Cadzand'. These territorial gains have often been dismissed as insignificant, or ignored altogether; the most thorough modern account of the Treaty of Dover notes in passing that Charles II was to receive 'two Dutch islands and a port', but does not even name them, let alone analyse their strategic significance.[35] However, these towns were not merely token prizes selected at random. In one sense, the transfer of the specified territory to Charles would simply have restored to him possessions held by his ancestors; Den Briel, Vlissingen and the nearby fort of Rammekens were the 'cautionary towns' ceded to Elizabeth I in 1585 as part of the price for her military support to the Dutch rebels, and they continued to be garrisoned by English troops

Fig. 4. A large English two-decker off Dover, flying Charles II's royal standard; painting by Jacob Knyff. Private collection.

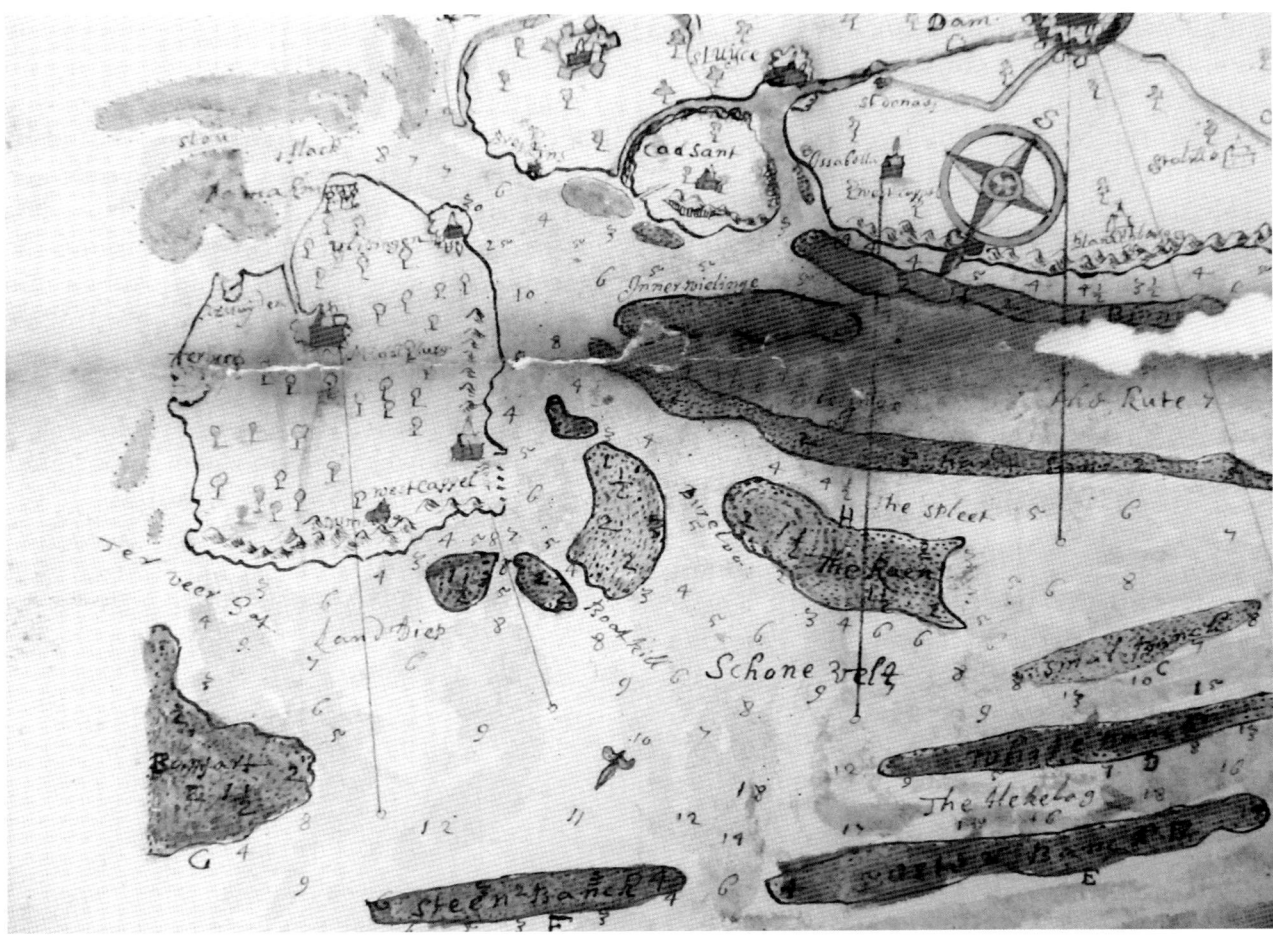

Fig. 5. A sketch map of the mouth of the Scheldt made in 1673 by the spy John Scott. It shows the island of Walcheren (including Vlissingen) and, on the opposite shore, Cadzand and Sluis. The National Archives.

until 1616, when James I sold them back to the province of Zeeland.[36] Den Briel and particularly Vlissingen were important centres for the Dutch navy's Zeeland admiralty; moreover Vlissingen was the home town of De Ruyter, leader of the attack on the Medway, who had actually spent the first nine years of his life growing up under the English flag. But regaining the cautionary towns in 1672-1673 would have meant far more than a symbolic restoration of a much older *status quo*, and a humiliating reverse for the Dutch to eclipse Chatham. Essentially, regaining control of the former 'cautionary towns' would at once have destroyed the finances of the Dutch navy, and would have handed to Charles II's England the power to open and close both the River Scheldt and the port of Antwerp.

Antwerp, once Europe's greatest port, fell to the Spanish in 1585 and thus lost its direct access to the sea, which now lay beyond a great swathe of Dutch territory. The Dutch created a barrier at their forts of Lillo and Liefkenshoek, ten miles downstream, and from 1585 onwards direct seaborne trade with Antwerp was prohibited. Instead, goods for inbound or outbound passage had to be transferred to other vessels at Lillo; after 1648 this maritime border moved to the anchorage between Vlissingen and Cadzand, the waterway assigned exclusively to Britain by the Treaty of Dover. The Dutch made a healthy profit out of the situation, charging for 'convoys' and 'licences' which granted the privilege of shipping to and from Antwerp; indeed, these two sources of income came to underpin the whole system of Dutch naval finance, as the funds were assigned solely to the five admiralties of the United Provinces.[37] During the 1660s France's increasingly aggressive designs on the Spanish Netherlands transformed the nature of 'the Scheldt question'. Louis XIV's finance and navy minister, Jean-Baptiste Colbert, noted how the decline of Antwerp had led directly to the rise of Amsterdam, and suggested that French control of Flanders might permit them to re-open Antwerp, with inevitably serious consequences for Dutch trade.[38] This possibility was understood within the English ministry. As Louis' army swept inexorably towards the Scheldt in August 1667, the Earl of Anglesey (then Treasurer of the Navy) informed the Duke of Ormonde that

> *the French success continues, and they offer Antwerp to open the Scheldt if they will yield, and to continue their ancient privilege, it's thought there will not long be peace between them [the French] and the Dutch.*[39]

By 1670, Louis XIV's ambitions were significantly greater. The scheme proposed at Dover envisaged much of the United Provinces, as well as Spanish Flanders, falling under French hegemony. Louis might legitimately have expected that, once Antwerp came under his control, he would restore it to the dominance in European maritime trade that it had once possessed, driving down Amsterdam in the process. Yet Louis effectively denied himself that opportunity by agreeing to cede control of both banks of the mouth of the Scheldt to his English cousin. If he did this deliberately, then surrendering the keys to Antwerp to Charles II was arguably almost as great a sacrifice and commitment on Louis' part as was the 'catholicising clause' on Charles's. On the other hand, there is some evidence to suggest that Louis' much-vaunted diplomats simply had no idea of the importance of the concessions they were making. When the leading minister Arnauld de Pomponne was briefed on the treaty terms in May 1671, following his return from an embassy to The Hague, he was horrified: he realised at once that the territories to be handed over to Charles II would give that monarch control of the Scheldt and a springboard for future interventions in Flanders and Holland. Hugues de Lionne, the old Secretary of State, merely laughed, perhaps a little nervously: 'To be perfectly frank, when we made the treaty it didn't occur to us that Middelburg and Flushing were on the island of Walcheren'.[40]

In any event, from the British viewpoint the prospect of acquiring such valuable territory on Dutch soil only entered the negotiating process between Britain and France relatively late in the day during the negotiations of 1669-1670.[41] Arguably, though, as the importance of the conversion to Catholicism declined in Charles II's estimation, so the prospect of regaining the cautionary towns, and transforming them into a strategically and economically lucrative fortress-colony, grew in significance. The king's sister Henrietta encouraged this train of thought. In September 1669 she wrote to Charles, stressing the importance of the towns that he would acquire:

> *Indeed what is there more glorious and more profitable than to extend the confines of your kingdom beyond the sea and to become supreme in commerce, which is what your people most passionately desire and what will probably never occur so long as the Republic of Holland exists?...[You should reserve] for yourself, in the division you will make, the most important maritime towns, whose commerce will depend entirely on the laws which you choose to impose upon them for the benefit of your kingdom and yourself...it is easy to see that the execution of the design which is being proposed to you would be the veritable foundation for your own greatness, because, having a pretext for keeping up troops outside your kingdom to protect your conquests, the thought alone of those troops, which for greater safety could be composed of foreigners and would be practically in sight of England, could keep it in check and render Parliament more amenable than it has been accustomed to be.*[42]

Thus by the Treaty of Dover Charles gained the more concrete, and more immediately achievable, concession, namely the Dutch territories; but if he was not sincere himself, and had no real intention of implementing the 'catholicising clause', then he may equally have doubted from the beginning the sincerity of Louis' commitment to hand over the cautionary towns.

In the short term, the emphasis on the annexation of territory in Zeeland, and the control of the Scheldt that accompanied it, helped to shape the elaborate diplomatic

Fig. 6. George Villiers, second Duke of Buckingham, the unwitting dupe of the Treaty of Dover. Trustees of the British Museum.

charade that was played out during the second half of 1670. During August and September a delegation headed by George Villiers, Duke of Buckingham, negotiated a second version of the treaty with the ministers of Louis XIV. This *traité simulé*, as it became known, was designed to win over to the war policy Buckingham and other ministers who were not privy to the secret version. The *traité simulé* naturally omitted the Catholicising clause, but it also contained one other important difference: in addition to the territories specified in the Dover treaty, Charles was also to receive the islands of Goeree and Voorne, which would effectively have given England control of the estuary of the Maas (and thus of the port of Rotterdam and naval base at Hellevoitsluis) as well as that of the Scheldt, effectively destroying yet another of the Dutch admiralties in the process.[43] Why did Louis concede more territory in the false treaty than in the real one? It is difficult to suggest explanations that are not unremittingly cynical. Firstly, of course, if Louis had no intention of handing over *any* land to Charles (in other words, if the promise of territory was just as empty as the Stuart king's promise to restore the mass) then it did not matter if he added a couple more islands for appearance's sake. The gesture might have been simply a ploy, designed to ensure that Buckingham and the others did not suspect that a secret treaty already existed.[44] It is also possible that the additional concession was part of the process of pandering to Buckingham's vanity, and to winning over the more reluctant members of the English court and ministry. In a nutshell, if Louis was genuine in proposing to hand over Walcheren and Cadzand to Charles, then adding Goeree and Voorn as well mattered relatively little; after all, he was simply giving away someone else's land, especially if his diplomats had only the haziest idea of where those lands actually were.[45] Moreover, the treaty without the catholicising clause looked very much like a sweeping acquisition of territory and control over Dutch trade routes by Britain without actually having to give very much in return, beyond committing its fleet and a very small army to a speedy and comprehensive annihilation of the Dutch state. As such, it would permit Buckingham to appear as the great diplomat who had managed to persuade the reluctant French to surrender such a great prize: in effect, most of the province of Zeeland, control of two great estuaries, and the power to open or close the port of Antwerp. In the *traité simulé*, Louis appeared to need Charles's assistance so badly that he gave him what he wanted without demanding serious concessions in return: in other words, the exact opposite of the truth, which was safely concealed in the secret document signed six months earlier.

This reading of events is supported by the progress of the abortive peace negotiations with the Dutch two years later, during the summer of 1672. Charles II was initially prepared to forego his claim to Cadzand and Sluis 'to gratify the French with yielding them to them, not to retard the making of the peace', but his ambassadors (Buckingham and Arlington) used their own initiative to insist on retaining the demand for the two towns, along with the inviolable demands for Vlissingen, Den Briel and a war indemnity of £500,000. As they argued,

what he is like to have besides [Cadzand and Sluis] will be of little value without them for the opening of the river to Antwerp, and consequently lessening the trade and value of Amsterdam, and this made us heretofore fight this battle so warmly with the ministers [of France], that perhaps they thought it fit to avoid the coming again to us in the same argument.[46]

Buckingham and Arlington clearly envisaged the re-opening of the Scheldt as an outcome of the proposed treaty, and it has been argued plausibly that they wished to retain the two towns on the left bank to prevent France closing it again to English trade in the future; but it is equally the case that the possession of Cadzand and Sluis, by giving Britain strong points on both banks of the river, would have made it easier to close the Scheldt to French and Flemish shipping (assuming that France followed up the conquest of the United Provinces by completing the annexation of Flanders, an outcome that was clearly envisaged in the Treaty of Dover's reference to 'new rights to the Spanish monarchy revert[ing] to the King of France').[47] After all, English trade had had almost a century to grow accustomed to the closure of the Scheldt and being deprived of access to Antwerp, but for France, the river's re-opening would have presented exciting new opportunities for its expanding overseas trades; France would have had rather more to lose if Charles II used his new lands to close the Scheldt once again, rather than vice-versa. In either case, though, the dispute over Cadzand and Sluis hardly suggests two allies who implicitly trusted each other. Although Charles II might still have wished to reassure Louis XIV of his good intentions by offering to abandon the claim to the two towns, Arlington and even the francophile Buckingham were clearly looking beyond the immediate outcome of the war to a future that seemed to suggest at least tentative grounds for Anglo-French hostility. Charles's comment to his Council at much the same time, 'the French will have us or Holland always with them, and if we take them not, Holland will have them', was hardly the sentiment of a monarch passionately committed to a French alliance and convinced of a 'special relationship' between himself and Louis XIV.[48]

Naval Preparations for War: the Greenhithe and Erith Dockyard Projects

Charles II and the Duke of York had begun the detailed ground work for their naval war with the Netherlands some time before their visit to Dover. On 11 May 1670 they attended a meeting with the Treasury commissioners, the Navy Board and the officers of the Ordnance. Samuel Pepys, Clerk of the Acts to the Navy Board, informed them that the total amount needed to pay off the navy's debt, to repair the fleet, to re-stock the storehouses, to finish the five large ships under construction and to keep fifty ships at sea in the summer of 1670 would be no less than £906,172 13s 1¾ d (rather more than half of which was the debt outstanding up to and including 1669, most of which was still the legacy of the second Dutch war).[49] Indeed, some of the navy's debts were even older: several dated back to 1660, and Thomas Chudleigh was still owed over £659 for 'workmanship' carried out during 1661.[50] A breakdown of the repairs required by each ship in the navy was drawn up a fortnight later, putting the total cost at £70,129.[51] The pace of the preparations increased within days of the Stuart brothers' return from Dover. On 14 June, they met once more with the Treasury commissioners and the Navy Board, presumably with Pepys again in attendance. Charles informed them that he wanted the debt of the navy reduced to such a level that it may be registered on the Wine Act [and] ordered that the preparations of the navy be gone in hand with all speed, also that in building the shipwrights never go beyond their dimensions on paper and have their timber in readiness in the yard and that the contrary be made penal.

Large supplies of hemp, the most expensive and in many respects the most essential of all naval stores, were to be purchased while prices remained at about half of what they had been during the second Dutch war (1664-1667).[52] Meetings such as those of 11 May and 14 June, with the king, the Lord High Admiral and all those involved in naval administration and naval funding in attendance, were very rare indeed, reflecting the seriousness of the agendas; Pepys recorded no such meetings during the entire nine-and-a-half years of the diary period (indeed, during that time there were only three meetings between Charles and the Navy Board alone). The king's message at the meeting of 14 June could hardly have been clearer. Restricting the size of the ships under construction, accelerating naval preparations and laying in large stocks of hemp were measures designed entirely to get a very large fleet to sea in the summer of 1671, and would help to ensure that the unprecedented programme of great first and second rate ships was completed before war broke out. Charles clearly envisaged fulfilling the Dover treaty terms and launching the war against the Dutch during 1671.

Confirmation of this comes from several pieces of evidence, seemingly insignificant in themselves, from the period between the summer of 1670 and the spring of 1671. For example, there is the categorical and urgent postscript added by Sir George Downing to a letter that he wrote to Alderman Edward Backwell, the crown's most important and trusted banker, and the joint Treasurers of the Navy (Sir Thomas Littleton and Sir Thomas Osborne), to attend at the Treasury on 2 August 1670 to discuss the provision of money for repairing ships: 'this meeting is by His Majesty's special direction; therefore pray fail not'.[53] Four contracts for various kinds of naval supplies were signed by the Navy Board during the first six months of 1670; 25 were then signed in August and September alone, the great majority being large orders for timber and plank.[54] A significant number of new ships were ordered. On 10 July James confirmed the order for two large galleys, one to be built at Genoa and the other at the Medici arsenal in Pisa. Both were intended for the defence of Tangier, ostensibly against the Barbary corsairs. In August, three new third rates were ordered from private yards, two in Hull and the third in Bristol.[55] The restored monarchy was normally reluctant to order ships under contract from merchant shipbuilders, and sought to avoid doing so as far as possible. Construction in private yards was more difficult to monitor than building 'in house' at the royal dockyards; it was believed (with some justification) that private builders might seek to cut corners and line their own pockets, and (with no justification whatsoever) that ships built in private yards were often inferior to those built in the king's.[56] Ordering the three new thirds from private contractors was a sure sign that Charles and James wished to keep their dockyards free to complete the work on the first and second rates that were already under construction, and to undertake the refurbishment of the existing fleet.[57] In September 1670 Charles and James gave orders for their 'fifty best ships' to be selected by the Navy Board and made ready for service in the spring of 1671. These were not to be the equivalent of the 50 ships that had been in service during 1670, most of which were fourth rates or smaller (only three larger ships, all thirds, were at sea in that year); the list eventually prepared by Pepys and his colleagues included all of the navy's first and second rates, including the new builds.[58] Brass guns were to be taken out of forts around the British Isles and re-cast for the 1671 fleet; meanwhile, the dockyards worked at full capacity through the winter of 1670-1671 to ready the fleet for sea.[59] In October Charles and James ordered a victualling declaration for 25,000 men to serve in the fleet during 1671. This was a covert declaration that war was imminent; 1670's 'fleet' had required victuals for only 5,000 men, but a declaration for 25,000 men would have been appropriate to man the 50 great ships demanded by the Treaty of Dover, and which were undoubtedly synonymous with the 'fifty best ships' selected for service in 1671.

The refurbishment of the existing fleet, the completion of the new 'great ships' and the proposed war against the Netherlands was to be accompanied by a vast infrastructure project that has been wholly ignored in all previous accounts of the Anglo-Dutch wars and the foreign policy of Charles II. In about July 1670, shortly after the conclusion of the Dover treaty, construction of an entirely new royal dockyard commenced at Greenhithe. The administration was soon receiving applications for posts as officers of the yard, a sure sign that the project was public knowledge and that its completion was taken for granted. But the ground at Greenhithe seems to have been unsatisfactory, and nothing more is heard of digging work at that site after the end of September 1670.[60] In November, though, John Tippetts, a commissioner of the navy, produced a report on potential dockyard sites along the Thames. He condemned the site at Greenhithe as too narrow and the ground there as too loose, soft and sandy to accommodate docks. By contrast, a site at Erith 'to the westward over against the church' had clay down to at least 19 ft (the deepest borehole that Tippetts dug) and ample space for the proposed facilities.[61] During the spring and summer of 1671 the dockyard project was focused exclusively on Erith, the site of an old Tudor dockyard; the 'great storehouse' from Henry VIII's time still stood there, possibly as a storage facility for the East India Company which made much use of the anchorage at Erith, and this

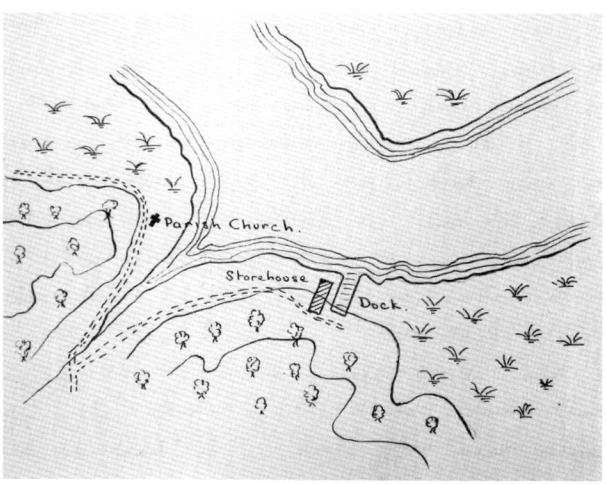

Fig. 7. Plan of the site of the Tudor dockyard at Erith, drawn in 1941 by Sergeant Donald Turner, RAF, who was killed in action in 1944. Bexley Archives.

building survived well into the nineteenth century.[62] The significance of Erith yard is that it was not to be a relatively small-scale forward repair base like Sheerness, the other new dockyard. This was nearing completion at the time but possessed only very limited facilities, including one small dry dock. By contrast, Erith yard would have been vast, outdoing even Chatham (the largest industrial site in the British Isles, if not Western Europe) in the scale and capabilities of its facilities. Chatham had one single and one double dry dock; Erith would have had two double docks, each 350 ft long.[63] Erith was to have both a wet dock, a facility that permitted enclosed above-water repair and laying-up facilities for the ships within it, and a mast dock. At the time, only Deptford possessed similar docks. Erith was to have 900 ft of wharfage; Chatham had significantly more, at about 1,300 ft, but that was due to and offset by its lack of a wet dock. Erith was to be equipped with several storehouses, the largest of which would be 1,000 ft long: more than twice the size of the 'great storehouse' at Deptford and the similar facility at Portsmouth. The whole was to be surrounded by a brick wall, twelve ft high, protected by six watchtowers. The yard was to cost £63,014; the site at Chatham, often regarded by contemporaries as one of the most prodigious spectacles in the kingdom, was worth only £45,000, and Erith yard would have cost about the same as the huge new palace at Winchester, commenced in 1682 as

Fig. 8. View of Erith and Long Reach, 1795. The site of the proposed dockyard of 1670 would probably have extended from near to the church towards the town in the right near distance. Trustees of the British Museum.

virtually a copy of Versailles.[64]

The case for having a yard at Erith was strong. The Thames became significantly shallower above that point, making it difficult or impossible for large ships to reach London (and the upstream dockyards at Deptford and Woolwich) in all states of the tide. There were other problems with the existing yards; they lacked space for expansion (especially Woolwich), and silting in the Medway, caused partly by the narrow arches of Rochester Bridge, made safe navigation to and from Chatham problematic.[65] Erith was significantly nearer to London than Chatham (and thus nearer to such essential naval services as

Fig. 9. Gravesend Blockhouse. Along with Tilbury Fort on the opposite shore and other batteries along the Thames, this would have guarded the approaches to Erith dockyard. Trustees of the British Museum.

victuals, timber and manpower), and may well have been intended to replace that demonstrably vulnerable yard in the longer term. Moreover, the new yard would be protected by fortifications far more impressive than those that had signally failed to deter the Dutch in 1667: the state-of-the-art Tilbury Fort, the building of which was commenced by Bernard De Gomme in September 1670 (contemporaneously with the start of work at Greenhithe), and the well-established blockhouse and batteries at Gravesend on the opposite bank of the river.[66] The design and location of Erith dockyard suggests that it was intended to provide both a safer laying-up facility for many of the navy's ships, rather than continuing to depend on the vulnerable Medway, and docking facilities that would permit a much more rapid 'turnaround' of battle repairs during wartime. Significantly, though, there was no mention of a new dockyard in the estimates of the naval debt and potential future costs drawn up before the signing of the secret Treaty of Dover; Greenhithe, and its far grander successor at Erith, only began to feature on Charles II's naval 'wish list' after the war strategy was agreed, and after promises of French subsidies and more substantial parliamentary funding were received during the summer and autumn of 1670.[67] However, Erith yard was not even begun by the time that war broke out in March 1672; realistically, it could never have been ready in time for any war fought in the early 1670s. The vast new royal dockyard at Erith would be a consequence, not a cause, of a final victory over the Dutch, and it would be ready to accommodate the much larger 'navy royal' that would be funded by such a victory; a navy that would have to defend the Stuarts' new territories in Zeeland, the Caribbean and the East Indies against the only European power capable of threatening them.[68] We will never know whether Erith was intended to replace Chatham or Deptford and/or Woolwich. The former is probably most likely, given its navigational difficulties and the vulnerability that had been exposed in

Fig. 10. View of Erith, 1750. The large building in the middle background, behind the large tree, is believed to be the storehouse of Henry VIII's dockyard; the building survived until demolished in the late nineteenth century. Bexley Archives.

1667. But it is at least possible that Charles II intended Erith to *supplement* all of the existing yards: paid for from the proceeds of the destruction of the Dutch state and the acquisition of both its riches and the Scheldt tolls, both it and a greatly expanded navy would ensure Britain's sovereignty of the seas under the future King Edgar.[69]

The Postponement of the War and the Possible Project for a Dockyard at Plymouth

If Charles seriously wished to launch the war that the Treaty of Dover had envisaged in 1671, he soon realised that practical politics and the unremitting tyranny of the purse strings would tell against him. Parliament, which had been prorogued from April, reconvened on 25 October. The Lord Keeper, the gouty old cavalier Sir Orlando Bridgeman, informed the members that Charles II had entered into various agreements with other powers, notably the Triple Alliance, but warned of a heightened state of tension among their neighbours:

> *he told them of the great preparations made by the French and Dutch and other[s of] their neighbours, which necessarily called for equal preparations here, the safety and honour of the kingdom being an island required it.*

Bridgeman made public the naval planning of the previous months by stating that a fleet of 50 ships, the exact number required under the terms of the Dover treaty, would be sent to sea in the following spring at a cost of £800,000. However, the income from the new wine duties was smaller than anticipated, and Parliament would need to rectify this if such a fleet was to be set out.[70] Discussions on supply dragged on through November. The MPs happily voted additional duties; both the customs and wine impositions and licences were handed to farmers who undertook to supply £600,000 per annum, £80,000 more than before, and these were to advance £150,000 by Michaelmas. On 10 December Charles attempted to focus the thinking of the MPs by claiming that they had still not voted enough to enable him to prepare defences against Louis XIV's anticipated expedition to Dunkirk in the spring, and although the Commons responded with some urgency, the actual returns still fell a long way short of the amount needed to fit out the fleet for either a fictitious mobilisation against Louis or a genuine campaign alongside him. The shortfall effectively ended any possibility of beginning the attack on the Dutch in 1671, though realistically, the chances of achieving that within such a short timescale were always remote; moreover, Louis XIV's other allies (the German states of Munster and Cologne) could not be ready in time, regardless of the political and naval preparations in Britain.[71] By the end of March 1671, Charles II was reflecting this reality by resolving to set out only 30 ships that summer.[72]

Charles II's essentially unsuccessful attempts to set out large fleets in both 1668 and 1671, his ambitious programme of new construction and his plans for the huge dockyard at Erith, can certainly be interpreted as aspects of the king's ambition to fight the Dutch once more and to avenge the humiliation of the Medway raid. However, they can also be seen as the Stuart contribution to the succession of grandiose political and military gestures that European monarchs made during the years 1667-1671, gestures that were intended both to impress and intimidate their neighbours. All sides were seeking to keep their options open, making secret treaties and then proposing agreements with third parties that directly contravened those treaties; one historian has argued that Charles's policies in this period

> *continued to edge cautiously towards a new Dutch war, while trying to keep an option of withdrawing from the whole scheme and attempting all the time to win as many concessions from as many nations as possible.*[73]

Charles was not the only ruler behaving in this way. In particular, Louis XIV's provocative actions during 1671 were hardly supportive of any attempts by Charles to prepare his nation for a public release of the news of an Anglo-French alliance and a new Dutch war. The 'Sun King's' long-anticipated arrival at Dunkirk in April,[74] accompanied by an army and a fleet, panicked the credulous; or, as one newsletter writer put it, 'most sober persons think we have as much cause (if not more) to fear him than the Dutch, for they *[the French]* have an army of 100,000 already raised, and a considerable fleet'. The Duke of Monmouth went over to Dunkirk to present his father's compliments to King Louis, but as an insurance policy, two troops of horse and ten companies of foot were sent into Kent to guard the coasts against a French attack launched from a town that had, after all, been British territory only nine years before.[75] The pitiful size of the force was at once a comment on the weakness of Charles II's armed forces and an obvious piece of 'window dressing' to placate public opinion, especially if Charles was certain that Louis' camp at Dunkirk was not directed against him.

The possibility of war *against* the French, if not imminently then at some point in the near future, was not wholly inconceivable. Even Charles II's sister Henrietta, the catalyst of the Dover treaty and both the sister-in-law and personal favourite of Louis XIV, had considered this possibility: in September 1669 she wrote to Charles,

I know well that there are some people who think that after France has increased her power by bringing about the downfall of Holland she would endeavour to take away from you your share of the conquests you will have made.

Although Henrietta denied the validity of this opinion, it was clearly shared by others, even after the uneasy allies went to war against the Netherlands.[76] The pamphleteer John Stubbe, a diehard advocate of war against the Dutch, contemplated the relative merits of fighting them or the French, and did not rule the possibility of a war against the latter in due course:

...it is at present best for the King of Great Britain to join his arms with those of France, leaving off the respects unto remote and perhaps only imaginary evils, that may never fall out...a war with Holland (upon account of prizes) is less expensive than a war with France: and it hath this further advantage, that it weakens our only competitor in naval strength; and the same maritime force which baffles Holland will secure us of the French amity, revenge the injuries we have received [my emphasis], and regain that honour which they have so villainously clouded and bereaved us of abroad...[77]

Stubbe's remarks beg the question of what might happen if France became 'our only competitor in naval strength', as it certainly would if the Dutch Republic was destroyed, and the evidence of Charles II's letters to Madame in 1668-1669, quoted above, suggest that he, too, felt at least a little of this concern. But if Charles was to contemplate war with France, either in the short or long term, then he would need a second new dockyard. The existing yards and the proposed facility at Erith were poorly sited to deal with a French fleet based at Brest, and there would clearly be a need for a base to the westward (as Charles's successors soon

Fig. 11. A Wenceslaus Hollar view of Plymouth and the Cattewater. Trustees of the British Museum.

realised when war against France finally broke out in 1689). In July 1671 a fleet of royal yachts carried Charles and his entourage to Plymouth, where the king inspected the new royal citadel, the existing harbour, and the river Hamoaze to the west.[78] Charles seems to have followed up this visit by commissioning Philip Lanyon, the navy agent at Plymouth, to investigate the possibility of upgrading the facilities there. Lanyon's report was sent directly to the King, rather than to his notional superiors the Duke of York and the Navy Board. It recommended the construction of a dry dock on the Cattewater, rather on the Hamoaze site where the dockyard was eventually built in the 1690s; Charles had inspected both sites during his visit. Lanyon also advocated the stationing at Plymouth of a squadron 'to keep Spain'; a concern for preserving Spanish neutrality would be one of the king's chief priorities during 1672-1673.[79]

Conclusion

In September 1671 Parliament was prorogued until October 1672, rather than April, as was originally intended. The revised timing was clearly intended to delay the new session to the end, rather than the beginning, of the putative military and naval campaign for that year, and that can only have been done with one intention in mind. If the scheme set down in the Dover treaty went according to plan, there was every chance that by the time it reconvened, Charles II would be able to present his suspicious MPs with a magnificent *fait accompli* that not even they could gainsay: the United Provinces crushed, Dutch trade and overseas possessions firmly in English hands,[80] Zeeland annexed, the Scheldt open or closed depending on the whim of the British king. In such circumstances, any inconvenient reminders of the 'catholicising clause' from Louis XIV could be safely ignored. Even if the parliamentarians were underwhelmed and ungrateful, a king with such a crushing victory behind him (and cushioned by the new income streams that such a victory would provide) might have no need of Parliament at all. One paper written in the summer of 1672 suggested that the destruction of the Dutch republic would be merely a stepping stone to taking over the Dutch East India Company's (VOC) possessions and trade in the East Indies, followed by the establishment of a monopoly in pepper and other spices, which would bring Charles II 'a very great revenue'. Such an accretion of riches might have been mere wishful thinking, partly because Louis XIV's so-called *escadre de Perse*, commanded by Jacob Blanquet de la Haye, had sailed from Rochefort for the Indian Ocean in March 1670 with orders to set up French fortified harbours in Ceylon and Indonesia – a far more direct and impressive intervention by a state navy in the East Indies than any contemplated by Charles II during the whole course of his reign. Although La Haye's squadron was never directly ordered to attack the positions of the VOC, the 'hidden agenda' of the squadron was well known, and Charles II refused to agree to it, presumably because of his own ambitions toward the Dutch trading posts in the east.[81] But if the inconvenient obstacle of the *escadre de Perse* could be overcome and English hegemony in the East Indies established, then taken together with the domination of the Scheldt and Maas and the subsidies from Louis XIV (for as long as they endured, which was presumably until the putative fate of the *escadre de Perse* became known), Charles II's financial dependence on Parliament would have been significantly reduced, if not ended outright.[82]

On 25 September, Charles moved to Newmarket for a pleasant three weeks of horse-racing and social calls.[83] During his absence, the apparent imminence of war with the Dutch finally became public knowledge. By the beginning of October, it was being said in London that a large fleet, commanded by an English admiral, would fight with the French against the Dutch in the following summer, and when Charles returned to Whitehall on 18 October, he had to order the Lord Mayor to suppress all such talk of an Anglo-French war against the Dutch.[84] One crucial piece of evidence suggests that this may not have been simply a piece of hypocritical authoritarianism on the king's part, and that he was genuinely not yet committed to going to war alongside France in the following spring. In the middle of October, the Navy Board requested the king's confirmation of the figure for the following year's victualling declaration. The 50 large ships required by the Dover treaty would have needed a declaration for 25,000 men, identical to the figure originally proposed for 1671. Instead, Charles refused to make the declaration, and early in November he even requested an estimate of the cost of fitting out the entire navy in the summer of 1672. These two actions suggest that as late as four months before the actual outbreak of war, Charles II still wished to have all his options open, and had not decided definitively to go to war alongside Louis XIV. His discussions with the Navy Board prove that he could have abandoned his plans for a large warfleet and simply victualled a token force of 5,000 men, as he had done in 1671, but he also clearly had in mind the possibility of deploying his entire navy unilaterally in the spring of 1672, regardless of the Treaty of Dover and regardless of anything that Louis might or might not do. Ultimately, of course, mobilising the entire fleet would give Charles insurance against a last-minute betrayal by Louis, but it would also have enabled him to 'think the unthinkable': to fight a war *against* France, the scenario that also seems to provide the best explanation for the strange chimera that was the royal dockyard at Erith.

References

[1] The account of the Medway attack that follows is based chiefly on P G Rogers, *The Dutch in the Medway*, 1970, 83-115; C J W Van Waning and A Van Der Moer, *Dese aengemaene tocht Chatham 1667*, Zutphen, 1981, especially pp. 39-66, 94-97.

[2] F Fox, *A Distant Storm: The Four Days' Battle of 1666*, Rotherfield, 1994, 346-347. I am grateful to Frank Fox for his comments on this paper.

[3] M Baker, 'Dutch War and French Peace', *Maritime South West*, 20, 2007, 63-64.

[4] See e.g. W De Britaine, *The Dutch Usurpation*, 1672, 27.

[5] There are few contemporary definitions or discussions of the concept of invasion. My interpretation is based chiefly on *England's Defence: A Treatise Concerning Invasion*, 1680, which was partly a recycling of some materials from the 'Armada Year'.

[6] Bod[liean Library, Oxford], Carte MS 47, fo. 166, Earl of Anglesey to Duke of Ormonde, 13 July 1667.

[7] Some sources gave her name as *Charles the Second*. Pepys remarked 'God send her better luck than the former': R Latham and W Matthews, eds, *The Diary of Samuel Pepys Diary*, 11 vols., 1970-1983, ix. 101, 3 March 1668; H[istorical] M[anuscripts] C[omission], *Le Fleming MSS*, 55.

[8] List of proposed Summer Guard, 21 January 1668: T[he] N[ational] A[rchives] SP 46/137/31.; C[alendar of] S[tate] P[apers], D[omestic], 1667-1668, 428-429; C H Hartmann, *Charles II and Madame*, 1934, 212, Charles to Henrietta, 14 June 1668.

[9] C[alendar of] S[tate] P[apers], V[enetian], 1666-1668, 187; HMC *Beaufort MSS*, 64; J D Davies, 'The Birth of the Imperial Navy? Aspects of English Naval Strategy, c.1650-1690', *Parameters of British Naval Power 1650-1850*, ed. M Duffy, Exeter, 1992, 19.

[10] When six first rates put to sea in the spring of 1672, it was observed that this was 'more than any King of England was ever Master of'; HMC, *Sixth Report*, 369. Cf E M Thompson, ed, *Correspondence of the Family of Hatton*, Camden Society, 1878, i. 59, 75.

[11] In this respect, it might be significant that on 31 March 1671 the new 100-gun first rate *Royal James* was launched at Portsmouth on the same tide as a new fifth rate frigate which Charles II named the *Phoenix* – perhaps signifying the 'rebirth' from the ashes of the *Royal James* that had been burned at Chatham.

[12] *CSPD 1668-1669*, 80, 88, 100, 122, 147, 167-168, 169, 174, 362, 407, 473, 581; *CSPD 1670*, 24, 70-71, 91, 170, 179, 196, 212-213, 532; *CSPD 1671*, 90, 127, 166, 181, 191, 193, 197, 203; Bod., Rawl[inson]. MS A195, fo. 62, list of ships building at Rochefort by 'JS', 15 October 1668; Rawl. MS A477, fos. 158, 161, 163, lists of French ships.

[13] It has been suggested that Louis' new navy was actually intended to divide the Habsburg forces and territories, and perhaps eventually to take possession of the Spanish Habsburgs' overseas possessions: J Glete, 'The sea power of Habsburg Spain and the development of European navies, 1500-1700', paper given to the conference *Guerra y Sociedad en la Monarquía Hispánica: Politica, Estrategia y Cultura en la Europa Moderna 1500-1700*, Madrid, 9-12 March 2005: http://www2.historia.su.se/personal/jan_glete/Glete-Sea_Power_Habsburg_Spain.pdf, accessed 10 September 2008. (I am grateful to Dr Peter Le Fevre for this reference.)

[14] Although Castle lived at Gosport, and seems to have been under the patronage of Anthony Deane, master shipwright at Portsmouth, he does not actually seem to have been employed in that yard: TNA, ADM 42/1215, Portsmouth extraordinary paybook.

[15] TNA, SP 78/125, fos. 23, 44, Castle to Williamson, 23 September and 9 October 1668.

[16] TNA, SP 78/125, fos. 57-60, 73-74, Castle to Williamson, 1 and 15 Nov. 1668, and MPF 1/325 (Castle's plan, extracted from SP 78/125). The mission, which seems to have ended in December 1668, is alluded to in *CSPD 1667-1668*, 591; *1668-1669*, 34, 90, 91. Thomas apparently died in the summer of 1669, as his will was proved on 28 July and his widow Hesther received £100 of royal bounty money in August: TNA, PROB 6/44. Fo. 84; C[alendar of] T[reasury] B[ooks], 1669-1672, 262, 270.

[17] Castle wrongly identified her as the *Dauphin Royal*, which was actually under construction at Toulon.

[18] TNA, SP 78/125, fo. 73, Castle to Williamson, 15 November 1668.

[19] *Ibid*, fos. 57-62.

[20] *Ibid*, fos. 64, 76.

[21] Hartmann, *Charles II and Madame*, 223-4, Charles to Henrietta, 2 September 1668.

[22] *Ibid*, 229, Charles to Henrietta, 20 January 1669.

[23] *Ibid*, 264-265.

[24] J A Lynn, *The Wars of Louis XIV 1667-1714*, 1999, 106.

[25] *Ibid*, 108; P Sonnino, *Louis XIV and the Origins of the Dutch War*, Cambridge, 1988, 9-21; H Hasquin, *Louis XIV face à l'Europe du Nord: l'absolutisme vaincu par les libertés*, Brussels, 2005, 121-128.

[26] M Lee, *The Cabal*, Urbana, 1965, 86-93; R Hutton, *Charles the Second*, 1989, 254-255; J Miller, *Charles II*, 1991, 144-146.

[27] W Troost, *William III, the Stadholder King*, Aldershot, 2005, 61.

[28] V C Kampmann, 'Der Englische Krone als 'Arbiter of Christendom'?: Die 'Balance of Europe' in der politischen Diskussion der späten Stuart-Ära (1660-1714)', *Historisches Jahrbuch*, 116, 1996, 321-351.

[29] In mid-December 1668 Charles sent his sister a cipher (something they had never needed before) and decided to prorogue Parliament from March to October 1669. For the timing, see Lee, *Cabal*, 101-103.

[30] Miller, *Charles II*, 160-166.

[31] R Hutton, 'The Making of the Secret Treaty of Dover, 1668-1670', *The Historical Journal*, 29, 1986, 300-302, 304-305.

[32] M Mignet, *Négociations relatives à la succession d'Espagne sous Louis XIV; ou, Correspondances, mémoires, et actes diplomatiques concernant les prétensions ... de la maison de Bourbon au trône d'Espagne. Accompagnés d'un texte historique*, Paris, 4 vols., 1835-

33 *CSPD 1670*, 263.

34 The full text is given in Mignet, *Négociations*, iii. 187-197. There is a detailed English summary in Hartmann, *Charles II and Madame*, 310-312.

35 Hutton, 'Treaty of Dover', 303.

36 J A Van Arkel, 'De inlossing der Engelse pandsteden in 1616', *Tijdschrift voor Geschiedenis*, 68, 1955, 59-69.

37 S T Bindoff, *The Scheldt Question to 1839*, 1945, 85-93, 110-111; R. Reitsma, 'Dutch Finance and the English Taxes in the Seventeenth Century', *Bestuurders en Geleerden*, ed. S Groenveld, M E H N Mout and I. Schöffer, Amsterdam, 1985, 108.

38 Bindoff, *Scheldt*, 126.

39 Bod., Carte MS 217, fo. 405, Anglesey to Ormonde, 31 August 1667.

40 Sonnino, *Dutch War*, 147.

41 See Miller, *Charles II*, 166-168.

42 Hartmann, *Charles II and Madame*, 278-280, Henrietta to Charles, 21 September 1669.

43 Bindoff, *Schedlt*, 127; Miller, *Charles II*, 178-179. The full text of the *traité simulé* is given in Mignet, *Negociations*, iii. 256-267.

44 Miller, *Charles II*, 179.

45 Sonnino, *Dutch War*, 128.

46 H T Colenbrander, ed, *Bescheiden uit vreemde archieven omtrent de groote Nederlandsche zeeoorlogen*, The Hague, 1919, ii. 169; Troost, *William III*, 81. Arlington and Buckingham got their way, and the final Anglo-French demands, drawn up on 6/16 July 1672, expanded the territorial claim to include the whole of the island of Walcheren: *ibid.*, 82.

47 Bindoff, *Scheldt*, 128.

48 TNA SP 104/177, fo. 84.

49 Printed in J R Tanner, ed, *Further Correspondence of Samuel Pepys* (1929), 266-267, and summarised in *CSPD 1670*, 205-206.

50 TNA PRO 30/32/2, 'List of His Majesty's Navy Debt' compiled in August 1670.

51 TNA SP 29/284/54.

52 *CTB, 1669-72*, 450.

53 *CTB, 1669-72*, 642.

54 *CSPD 1670*, 608.

55 TNA ADM 106/20/39, 61, 85, James to Navy Board, 10 July, 4 August, 18 August 1670.

56 J D Davies, *Pepys's Navy: Ships, Men and Warfare, 1649-89*, Barnsley, 2008, 68-69.

57 This was explicitly the case with the three thirds and six fourths ordered in September 1671. Additionally, a second rate was ordered from Portsmouth Dockyard in April 1671: TNA ADM 106/21/76, 273, 307, James to Navy Board, 20 April, 29 August, 16 September 1671.

58 TNA ADM 106/20/141, James to Navy Board, 25 September 1670; SP 29/285/298, list of 50 ships proposed for service in 1671; *CSPD 1670*, 498.

59 *CSPD 1670*, 556, 570; *CSPD 1671*, 91, 149.

60 *CSPD 1670*, 326, 429, 438, 444, 459, 469.

61 B[ritish] L[ibrary], Addit[ional] MS 9307, fo. 98, Tippetts to Navy Board, 1 November 1670.

62 C J Smith, *Erith: Its Natural, Civil and Ecclesiastical History*, 1873, 61; J J Wilkson, *History of Erith*, 1879, 50; J Harris, *The Parish of Erith in Ancient and Modern Times*, 1885, 72. I am grateful to Oliver Wooller, community archivist with Bexley Council, and Dr Clifford Pereira for their unstintingly generous provision of information about the Tudor dockyard at Erith and the local topography, and to Simon McKeon and the staff of Bexley Archives for facilitating my photographing of the 1750 Boydell print.

63 The evidence is incomplete, but it seems likely that only one 350-foot double dock was to be built at Greenhithe: *CSPD 1670*, 429. Although the French had included a stone dry dock within their new dockyard at Rochefort, commenced in 1666 and largely complete by 1669, this example was not emulated at Greenhithe or Erith, which were to have traditional timber-lined docks.

64 M Oppenheim, 'The Royal Dockyards', *The Victoria History of Kent*, 1908, ii. 355; D C Coleman, 'Naval Dockyards Under the Later Stuarts', *Economic History Review*, second series, 6, 1953, 139n. Oppenheim and Coleman seem to be the only historians to have noticed the 1670 dockyard scheme, albeit very briefly, but neither put it in the political and strategic contexts of the time. Moreover, Oppenheim seems to have assumed that Greenhithe and Erith were the same place, while Coleman assumed that two separate dockyards were to be built at the two locations.

65 Oppenheim, 'Royal Dockyards', 355; P MacDougall, 'The Abortive Plan for Northfleet Naval Dockyard During the Napoleonic Wars', *Archaeologia Cantiana*, 120, 2000, 151-152.

66 A Saunders, *Fortress Builder: Bernard De Gomme, Charles II's Military Engineer*, Exeter, 2004, 192-198; V Smith, 'Trinity Fort and the defences of the second Anglo-Dutch war at Gravesend in 1667', *Archaeologia Cantiana*, 114, 1995 for 1994, 39-50; V Smith, *The Gravesend Blockhouse*, 2000.

67 Indeed, no funding for Erith Dockyard was ever provided from 'official' sources, such as exchequer orders, suggesting that the project was to be paid for by the less public income streams that Charles II either received from Louis XIV or expected to receive from a victory over the Dutch.

68 It may be objected that a dockyard at Erith would have been of little use in any future conflict with the French. However, that would not have been the case if France effectively controlled most or all of Flanders and the Netherlands. The similar strategic situation after 1797 led to a project to build a vast new dockyard between Northfleet and Greenhithe: MacDougall, 'Northfleet Naval Dockyard', 149-168.

69 As it was, Edgar, Duke of Cambridge, died on 7 October 1671, aged four.

70 BL Addit. MS 36,916, fo. 193, newsletter to Sir Willoughby Aston, 25 October 1670.

71 BL Addit. MS 36,916, fos. 193-199, newsletters to Sir Willoughby Aston, 1, 8, 15, 22, 29 November, 6 December 1670. For the parliamentary debates over supply, see Lee, *The Cabal*, 143-144; J Spurr, *England in the 1670s*, 2000, 16-18; Miller, *Charles II*, 182-186. For the postponement of the 1671 campaign, see Sonnino, *Dutch War*, 122-124.

72 BL Addit. MS 36,916, fo. 211, newsletter to Sir Willoughby Aston, 28 March 1671.

73 Hutton, *Charles the Second*, 273.

74 Charles II was informed of Louis' intention to visit Dunkirk on 7 December 1670: HMC, *Seventh Report*, 489.

75 BL Addit. MS 36,916, fos. 218, 219, newsletters to Sir Willoughby Aston, 11, 18 April 1671; *CSPD 1671*, 181, 183.

76 Hartmann, *Charles II and Madame*, 278, Henrietta to Charles, 21 September 1669.

77 J Stubbe, *A Further Justification of the Present War*, 1673, 18, 19.

78 J J Beckerlegge, 'Charles II's visits to Plymouth', *Devonshire Association Report and Transactions*, 100, 1968, 220-221. Charles had originally intended an East Coast voyage, to Yarmouth and Hull: *CSPD 1671*, 170.

79 Bod., Rawl. MS A174, fo. 352, Lanyon to Charles II, undated but probably winter 1671-1672.

80 Certainly not 'British', as the exclusion of the Scots and Irish under the provisions of the Navigation Acts would presumably have remained in force.

81 G J Ames, 'Colbert's Indian Ocean Strategy of 1664-1674: A Reappraisal', *French Historical Studies*, 16, 1990, 536-559 (especially 545-554).

82 TNA, PRO 30/24/4, fo. 225.

83 BL Addit. MS 36,916, fos. 231-232, newsletters to Sir Willoughby Aston, 25 September, 28 October 1671.

84 *Ibid.*

Dr J D Davies *is the chairman of the Naval Dockyards Society. A Fellow of the Royal Historical Society, he has also served as a vice-president of the Navy Records Society and a member of the council of the Society for Nautical Research. His website is www.jddavies.com*